NASA

APOLLO SPACECRAFT

LUNAR EXCURSION MODULE

NEWS REFERENCE

APOLLO

Man's centuries-old dream of exploring the moon has been accomplished. Preparations for these half-million mile round trips included agencies from virtually all 50 states and the diverse talents of more than 300,000 people in government, industry, and the educational community.

All of them were well aware of their pioneering responsibility. And no matter where the scene shifted - from NASA installations in Washington, at Huntsville, Houston, or Goddard, to facilities at Bethpage or Cape Kennedy - all shared the same objectives.

However, in contrast to the initial moon landings which tested man and equipment, we are now embarking on more scientifically oriented explorations - resulting in longer lunar stay missions. These missions will provide the opportunity for a more comprehensive scientific study of the earth-moon relationship. A study that will expand our knowledge and understanding for the benefit of all mankind.

The goal is never out of sight. On a clear night it still beckons overhead - the MOON.

The Apollo Spacecraft News Reference has been prepared by Public Affairs, Space, at Grumman Aerospace Corporation, Bethpage, New York, in cooperation with the National Aeronautics and Space Administration, Manned Spacecraft Center, Houston, Texas.

All inquiries regarding the Apollo Lunar Module should be directed to:

Public Affairs, Space
Grumman Aerospace Corporation
Bethpage, New York 11714

CONTENTS

MISSION DESCRIPTION

A typical mission of the Lunar Module (LM) begins shortly after its separation from the coupled, orbiting Command/Service Module, continues through lunar descent, lunar stay, lunar ascent, and ends at rendezvous with the orbiting Command/Service Module before the return to earth. The LM mission is part of the overall Apollo Mission, the objective of which is to land two astronauts and scientific equipment on the moon, and return them safely to earth.

The three-stage, Saturn V launch vehicle will be used to boost the Apollo spacecraft into earth orbit and to provide the thrust necessary to propel it into its translunar path. Once on course to the moon, the third, and final propellant stage of the Saturn V is jettisoned, and the spacecraft (consisting of the Command, Service, and Lunar Modules) continues its journey toward a lunar orbit.

R-106

Upon approach to the moon, the re-ignitable propulsion system contained in the Service Module inserts the spacecraft into orbit above the lunar surface. Once orbit is achieved, two of the three astronauts in the Apollo team transfer from the Command Module, to the Lunar Module. A thorough checkout of the Lunar Module systems is then performed.

At a predetermined point in the Lunar orbit, the LM separates from the Command/Service Module which remains in lunar orbit awaiting the return of the LM at the end of the mission's rendezvous maneuver.

GRUMMAN

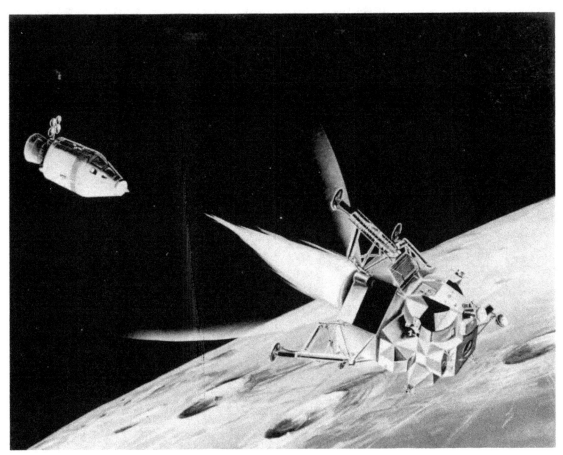

R-107

By igniting the throttleable descent engine (contained in the LM's descent stage), the LM's velocity is reduced and the spacecraft begins its trajectory toward the proposed landing site.

The LM's descent is automatically controlled to an altitude of a few hundred feet by its Guidance, Navigation, and Control Subsystem. During the final landing phase, the two man crew selects a favorable landing site and, by manual control of the reaction control system jets (clustered at the four corners of the LM ascent stage) and the variable thrust descent engine, maneuvers the craft to the correct attitude over the landing site and guides it to a gentle touchdown on the moon.

R-108

GRUMMAN

R-159

R-109A

Following touchdown on the moon, the LM crew checks all systems to determine whether damage was incurred during the landing, and to ensure that the systems will perform the functions required for a successful departure and rendezvous. All equipment not essential for the lunar stay is turned off.

With the LM secured for the lunar stay, both astronauts don their portable life support system, depressurize the LM and the Commander leaves the module to inspect the exterior for damage. As he descends the LM ladder, he actuates the mechanism which deploys the modularized equipment stowage assembly (MESA) containing a TV camera and other equipment used on the lunar surface. The Commander is followed shortly by the LM pilot.

R-160

R-162

R-161

R-163

Geological explorations are made by both astronauts, after which they return to the LM to replenish their portable life support system with on-board supplies.

During their explorations, the astronauts take many photographs, collect specimens, activate experiments, and transmit verbal reports on observations to the earth.

The lunar stay lasts for about 60 hours.

GRUMMAN

R-111A

R-164

When the lunar stay is completed, the crew prepares the LM for launch and ascent to rendezvous with the orbiting Command/Service Module.

Preparation consists of a recheck of all subsystems, and the computation of relative position information for the rendezvous.

When the orbiting Command/Service Module is in the proper position overhead, the LM ascent engine is ignited for launch. The descent stage serves as a launching platform and remains on the moon's surface. Later, the ascent engine inserts the LM into a transfer orbit. Mid-course correction and rendezvous maneuvers are accomplished with the reaction control system jets.

R-112

When the LM is approximately 500 feet from the Command/Service Module, the LM Commander manually maneuvers the module to a docking attitude and increases or decreases the rate of closure until complete docking is accomplished.

Once the coupling process is complete, the two-man LM crew prepares to transfer to the Command Module and rejoin the third member of the Apollo team. Pressures between the modules are equalized, LM subsystems are turned off, and scientific equipment and collected specimens are passed into the Command Module. When the transfer is complete, the LM is jettisoned in lunar orbit and left behind. This concludes the role of the LM in the Apollo mission.

FINAL APOLLO MISSION PHASE

Following rendezvous and jettisoning of the LM, preparations are made for the return journey to earth. A checkout of the Command/Service Module's systems and computation of the transearth course occur just before firing the Service Module's engine, which provides thrust for the return trip.

As the Command/Service Module nears earth, the Service Module is jettisoned, and the Command Module — bearing the Apollo team's three astronauts — is reoriented for re-entry and final parachute descent to earth landing, thus completing the week-long lunar mission.

APOLLO SPACECRAFT

The Apollo spacecraft comprises the lunar module, the command module, the service module, the spacecraft-lunar module adapter, and the launch escape system. The five parts, 82 feet tall when assembled, are carried atop the launch vehicle.

After the launch escape system and the launch vehicle have been jettisoned, the three modules remain to form the basic spacecraft. The command module carries the three astronauts to and from lunar orbit. The service module contains the propulsion system that propels the spacecraft during the translunar and transearth flights. The lunar module carries two astronauts, the Commander and the Lunar Module Pilot, to and from the moon, and serves as the base of operations during the lunar stay.

LUNAR MODULE

The lunar module will be operated in the vacuum of space; there was no need, therefore, for it to have the aerodynamic symmetry of the command module. The lunar module outer configuration was dictated only by the requirements of component location; cabin configuration was designed to provide a near perfect operating environment for the astronauts.

The LM consists of an ascent stage and a descent stage. Both stages perform as a single unit during separation from the CM, lunar descent, and lunar stay. The descent stage serves as a launching platform from which the ascent stage lifts off from the lunar landing site. The ascent stage operates independently during the lunar ascent, rendezvous, and docking phases of the mission.

ASCENT STAGE

The ascent stage is the control center of the LM; it is manned by the Commander, who occupies the left flight station, and the Lunar Module Pilot, who occupies the right flight station. The astronauts transfer to the ascent stage, through the docking tunnel, after the LM has docked with the CM and both have attained lunar orbit. The ascent stage comprises three major areas: crew compartment, midsection, and aft equipment bay. The cabin, comprising the crew compartment and midsection, has an overall volume of 235 cubic feet.

Because the LM is operated in either the weightlessness of space or in lunar gravity, the cabin contains harness-like restraint equipment rather than the foldable couches provided in the CM. The restraints allow the astronauts sufficient freedom of movement to operate all LM controls while in a relatively upright position.

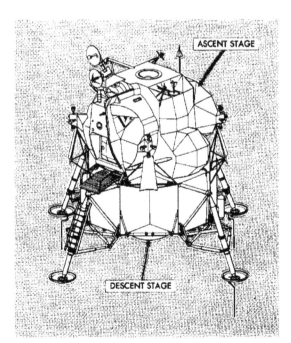

R-1A

Lunar Module

DESCENT STAGE

The descent stage is the unmanned portion of the LM; it represents approximately two-thirds of the weight of the LM at the earth-launch phase. In addition to containing the descent propulsion section, the descent stage is designed to:

Support the ascent stage

Provide storage to support the scientific equipment and the lunar roving vehicle used on the lunar surface.

Provide for attachment of the landing gear

Serve as the ascent stage launching platform

The descent stage is separated into five equally sized compartments that contain descent propulsion section components. The center compartment houses the descent engine; fuel, oxidizer, and water tanks are distributed in the remaining four compartments.

COMMAND MODULE

Dimensions

Height	10 ft 7 in.
Diameter	12 ft 10 in.
Weight (including crew)	13,000 lb
Weight (splashdown)	11,700 lb

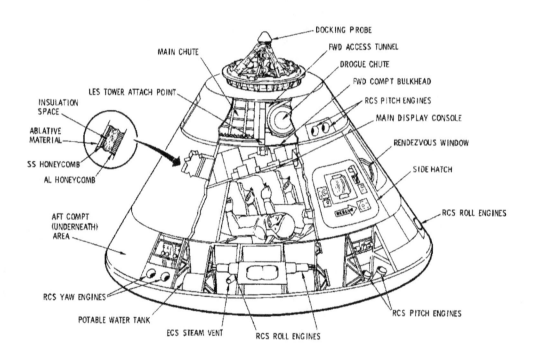

Command Module

P-47

GRUMMAN

Propellant	245 lb.

Reaction control subsystem
(fuel—monomethylhydrazine;
oxidizer—nitrogen tetroxide)

Function

The command module is the control center and living quarters for most of the lunar mission; one man will spend the entire mission in it and the other two will leave it only during the lunar landing. It is the only part of the spacecraft recovered at the end of the mission.

Major Subsystems

Communications
Earth landing
Electrical power
Environmental control
Guidance and navigation
Launch escape
Reaction control
Stabilization and control
Thermal protection (heat shields)

The CM is divided into three compartments: forward, crew, and aft. The forward compartment is the relatively small area at the apex of the module, the crew compartment occupies most of the center section of the structure, and the aft compartment is another relatively small area around the periphery of the module near the base.

During boost and entry the CM is oriented so that its aft section is down, like an automobile resting on its rear bumper. In this position the astronauts are on their backs; the couches are installed so that the astronauts face the apex of the module. In the weightlessness of space the orientation of the craft would make little difference except in maneuvers like docking, where the craft is moved forward so that the probe at the CM's apex engages the drogue on the LM. Generally, however, the module will be oriented in space so that its apex is forward.

Crewmen will spend much of their time on their couches, but they can leave them and move around. With the seat portion of the center couch folded, two astronauts can stand at the same time. The astronauts will sleep in two sleeping bags which are mounted beneath the left and right couches. The sleeping bags attach to the CM structure and have restraints so that a crewman can sleep either in or out of his space suit.

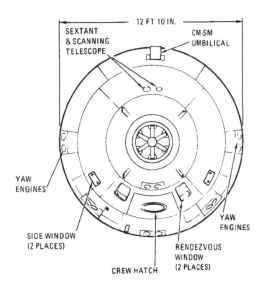

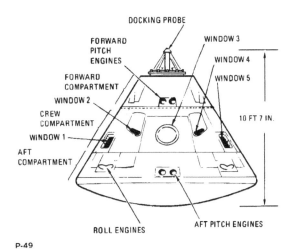

P-49

CM General Arrangement

Food, water, clothing, waste management, and other equipment are packed into bays which line the walls of the craft. The cabin normally will be pressurized to about 5 pounds per square inch (about a third of sea level pressure) and the temperature will controlled at about 75° F. The pressurization and controlled atmosphere will enable the three crewmen to spend much of their time out of their suits. They will be in their space suits, however, during critical phases of the mission such as launch, docking, and crew transfer.

The astronaut in the left-hand couch is the spacecraft commander. In addition to the duties of command, he will normally operate the spacecraft's flight controls. The astronaut in the center couch is the CM pilot; his principal task is guidance and navigation, although he also will fly the spacecraft at times. On the lunar mission, he is the astronaut who will remain in the CM while the other two descend to the surface of the moon. The astronaut in the right-hand couch is the LM pilot and his principal task is management of spacecraft subsystems.

Although each has specific duties, any of the astronauts can take over the duties of another. The command module has been designed so that one astronaut can return it safely to earth.

SERVICE MODULE

Dimensions

Height	24 ft 2 in.
Diameter	12 ft 10 in.
Weight (loaded)	54,000 lb.
Weight (dry)	13,475 lb.

Propellant

SPS fuel	15,690 lb.
SPS oxidizer	25,106 lb.
RCS	1,342 lb.

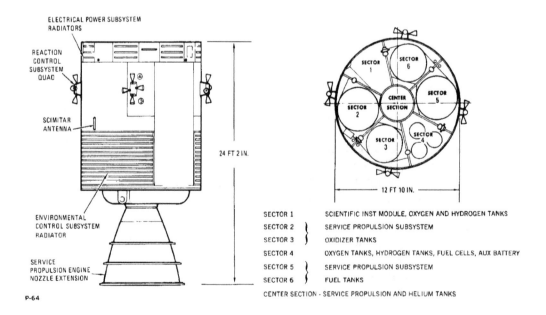

SECTOR 1   SCIENTIFIC INST MODULE, OXYGEN AND HYDROGEN TANKS
SECTOR 2 } SERVICE PROPULSION SUBSYSTEM
SECTOR 3 } OXIDIZER TANKS
SECTOR 4   OXYGEN TANKS, HYDROGEN TANKS, FUEL CELLS, AUX BATTERY
SECTOR 5 } SERVICE PROPULSION SUBSYSTEM
SECTOR 6 } FUEL TANKS
CENTER SECTION - SERVICE PROPULSION AND HELIUM TANKS

Service Module

Function

The service module contains the main spacecraft propulsion system and supplies most of the spacecraft's consumables (oxygen, water, propellant, hydrogen). It is not manned. The service module remains attached to the command module until just before entry, when it is jettisoned and is destroyed during entry.

Major Subsystems

Electrical power
Environmental control
Reaction control
Service propulsion
Telecommunications
Scientific Instrument module

The service module is a cylindrical structure which serves as a storehouse of critical subsystems and supplies for almost the entire lunar mission. It is attached to the command module from launch until just before earth atmosphere entry.

The service module contains the spacecraft's main propulsion engine, which is used to brake the spacecraft and put it into orbit around the moon and to send it on the homeward journey from the moon. The engine also is used to correct the spacecraft's course on both the trips to and from the moon.

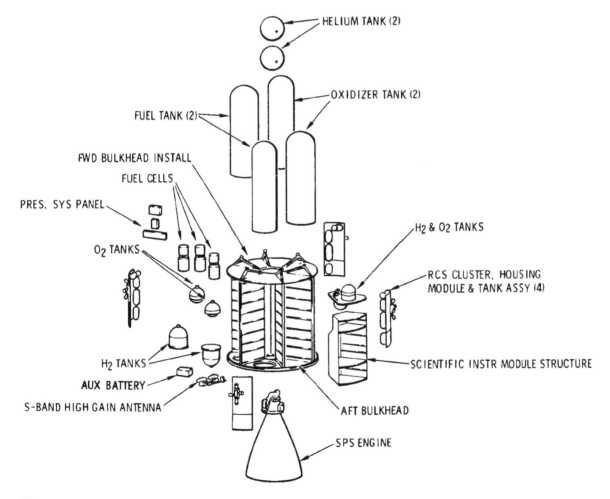

HELIUM TANK (2)

OXIDIZER TANK (2)

FUEL TANK (2)

FWD BULKHEAD INSTALL

FUEL CELLS

PRES. SYS PANEL

O_2 TANKS

H_2 & O_2 TANKS

RCS CLUSTER, HOUSING MODULE & TANK ASSY (4)

H_2 TANKS

AUX BATTERY

S-BAND HIGH GAIN ANTENNA

SCIENTIFIC INSTR MODULE STRUCTURE

AFT BULKHEAD

SPS ENGINE

Main Components of SM

Besides the service propulsion engine and its propellant and helium tanks, the service module contains a major portion of the electrical power, environmental control, and reaction control subsystems, and a small portion of the communications subsystem.

It is strictly a servicing unit of the spacecraft, but it is more than twice as long and more than four times as heavy as the manned command module. About 75 percent of the service module's weight is in propellant for the service propulsion engine.

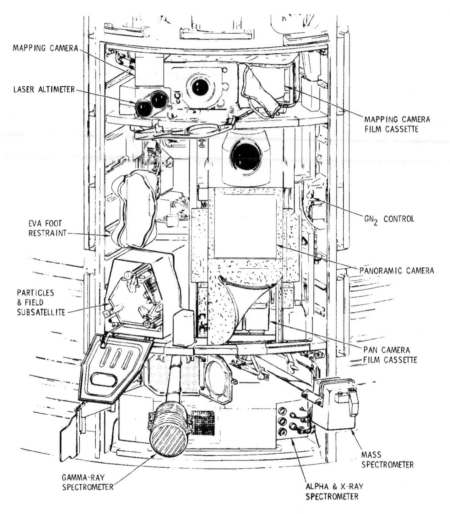

MAPPING CAMERA

LASER ALTIMETER

MAPPING CAMERA FILM CASSETTE

EVA FOOT RESTRAINT

GN$_2$ CONTROL

PARTICLES & FIELD SUBSATELLITE

PANORAMIC CAMERA

PAN CAMERA FILM CASSETTE

MASS SPECTROMETER

GAMMA-RAY SPECTROMETER

ALPHA & X-RAY SPECTROMETER

SM-2A-2217D

J-Mission SIM Bay, Apollo 15-16

Information in this section relative to the Command and Service Module was provided by North American Rockwell Corporation, Space Division. Complete details on the Command and Service Modules are contained in North American's Apollo Spacecraft News Reference.

LUNAR MODULE

QUICK REFERENCE DATA

DIMENSIONS

  LM:
    Height   22 ft. 11 in. (legs extended)
    Diameter   31 ft. (diagonally across extended landing gear)

  Ascent stage:
    Height   12 ft. 4 in.
    Diameter   14 ft. 1 in.

  Descent stage:
    Height   10 ft. 7 in.
    Diameter   13 ft. 10 in.

GENERAL

  Vehicle weight:
    Earth launch (with crew and propellant)   36,100 lb. (approx.)
    LM (dry)   10,800 lb. (approx.)
    Ascent stage (dry)   4,700 lb. (approx.)
    Descent stage (dry)   6,100 lb. (approx.)

  Propellant weight:
    Ascent stage   5,200 lb. (approx.)
    Descent stage   19,500 lb. (approx.)
    RCS   600 lb. (approx.)

  Pressurized volume   235 cu. ft.
  Habitable volume   160 cu. ft.
  Cabin temperature   75° F
  Cabin pressure   4.8 ± 0.2 psia

  Batteries:
    Height   3.03 inches
    Width   2.75 inches
    Length   6.78 inches
    Weight (each, filled)   135 pounds

  Electrical requirements:
    Inputs
      From Electrical Power Subsystem (Commander's   28 volts dc
      and LM Pilot:s buses)
      From Ascent engine latching device of control   28 volts dc
      electronics section
      From Explosive Devices batteries (systems A   37.1 volts dc (open-circuit voltage)
      and B)   35.0 volts dc (minimum)
      From descent engine control assembly   28 volts dc

    Outputs
      To initiators (in cartridge assemblies)   3.5 amperes for 10 milliseconds (minimum)
      Explosive Devices relay boxes   7.5 to 15.0 amperes dc (for at least 10
                                  milliseconds)

The NASA/Grumman Apollo Lunar Module (LM) after descending to the lunar surface from lunar orbit, provides a base from which the astronauts explore the landing site and enables the astronauts to take off from the lunar surface to rendezvous and dock with the orbiting Command and Service Modules (CSM). The LM consists of an ascent stage and a descent stage. Both stages function as a single unit during separation from the CM, lunar descent, and lunar stay. The descent stage serves as a launching platform from which the ascent stage lifts off from the lunar surface. The ascent stage operates independently during the lunar ascent, rendezvous, and docking phase of the Apollo mission.

The ascent and descent stages are joined by four interstage fittings that are explosively severed at staging. Subsystem lines and umbilicals required for subsystem continuity between the stages are either explosively severed or automatically disconnected when the stages are separated.

ASCENT STAGE

The ascent stage, control center of the LM, is comprised of three main areas: crew compartment, midsection, and equipment bay.

The crew compartment and midsection make up the cabin, which has an overall volume of 235 cubic feet. The basic structure is primarily aluminum alloy; titanium is used for fittings and fasteners. Aircraft-type construction methods are used. Skin and web panels are chemically milled to reduce weight. Mechanical fasteners join the major structural assemblies with epoxy as a sealant. Structural members are fusion welded wherever possible, to minimize cabin air pressurization leaks. The basic structure includes supports for thrust control engine clusters and various antennas. The entire basic structure is enveloped by thermal insulation and a micrometeoroid shield.

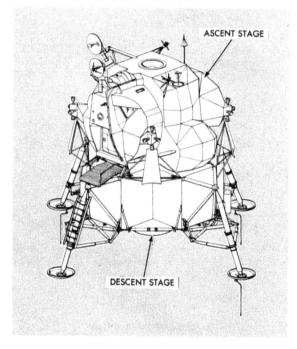

R-2A

LM Configuration

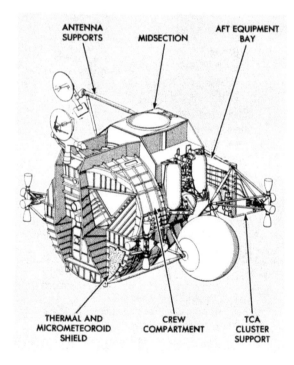

R-3

Ascent Stage Structure

The ascent stage is designed to:

Provide a controlled environment for the two astronauts while separated from the CSM.

Provide required visibility for lunar landing, stay, and ascent; and for rendezvous and docking with the CM.

Provide for astronaut and equipment transfer between the LM and CM and between the LM and the lunar surface.

Protect the astronauts and the equipment from micrometeoroid penetration.

CREW COMPARTMENT

The crew compartment is the frontal area of the ascent stage; it is cylindrical (92 inches in diameter and 42 inches deep). The Commander's flight station is at the left; the LM Pilot's at the right. The flight-station centerlines are 44 inches apart. For maximum downward vision the upper part of the compartment is constructed to extend forward of the lower portion. The area has control and display panels, body restraints, landing aids, a front window for each astronaut, a docking window above the Commander's station and other accessory equipment. Each flight station has an attitude controller, a thrust/translation controller, and adjustable armrests. There is a hatch in the front face assembly of the compartment.

A portable life support system (PLSS) donning station is behind the optical alignment station. Attachment points for an S-band in-flight antenna are provided on the front face assembly and for a rendezvous radar antenna on the upper structural beams of the crew compartment.

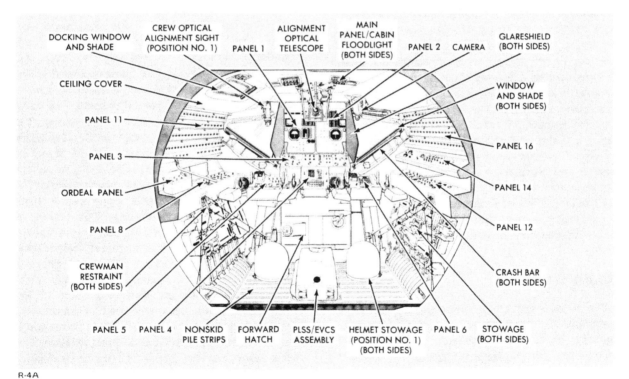

R-4A

Crew Compartment Interior

The crew compartment deck (flight station floor) measures approximately 36 by 55 inches. Nonflammable Velcro pile is bonded to the decks' top surface; a hooked Velcro on the soles of the astronauts' boots provides a restraining force to hold the astronaut to the deck during zero-g flight. Handgrips, aligned with the forward hatch and recessed in the deck, aid ingress and egress.

The control and display panels contain all devices necessary to control, monitor, and observe subsystems performance. The arrangement of the panels permits either astronaut to fly the LM to the CSM. All panels are canted to facilitate viewing. Six of the panels are in front of the flight stations. The upper two panels — one inboard of each flight station — are at eye level. These panels are shock mounted to dampen vibrations. The next two lower panels are centered between the flight stations to enable sharing of the control functions. One of the remaining two front panels is in front of each flight station, at waist height. The Commander's panel contains lighting, mission timer, engine, and thrust chamber controls. The LM Pilot's panel has abort guidance subsystem controls. To the left of the Commander's station are three panels: a five-tier circuit breaker panel at the top, an explosive devices and communications audio control panel, and an earth and lunar orbital rate display panel. To the right of the LM Pilot's station are three panels: the uppermost is a four-tier circuit breaker panel, the center panel contains controls and displays for electrical power, and the bottom panel contains communications controls and displays. The circuit breaker panels are canted to the line of sight so that the white band on each circuit breaker can be seen when the breakers are open.

FORWARD HATCH

The forward hatch is in the front face assembly, just below the lower display panels. The hatch is used for transfer of astronauts and equipment between the LM and lunar surface, or for in-flight extravehicular activity (EVA) while docked with the CM. The hatch is approximately 32 inches square; it is hinged to swing inboard when opened. A cam latch assembly holds the hatch in the closed position; the assembly forces a lip, around the outer circumference of the hatch, into a preloaded elastomeric silicone compound seal secured to the LM structure. Cabin pressurization forces the hatch lip further into the seal, ensuring a pressure-tight contact. A handle is provided on both sides of the hatch, for latch operation. To open the hatch, the cabin must be completely depressurized by opening a cabin relief and dump valve on the hatch. When the cabin is completely depressurized, the hatch can be opened by rotating the latch handle. The cabin relief and dump valve can also be operated from outside the LM. Quick-release pins in the latch plate and hinges may be pulled from inside the LM to open the hatch in an emergency.

WINDOWS

The two triangular windows in the front face assembly each have approximately 2 square feet of viewing area; they are canted down to the side to permit adequate peripheral and downward visibility. The docking window above the Commander's flight station has approximately 80 square inches of viewing area and provides visibility for docking maneuvers. All three windows consist of two separated panes, vented to space. The outer pane is of low-strength, annealed material that inhibits micrometeoroid penetration. The outer surface of this pane is coated with 59 layers of blue-red thermal control, metallic oxide, to reduce infrared and ultraviolet light transmission. The inner surface of the outer pane has a high-efficiency, antireflective coating. This coating is also a metallic oxide, which reduces the mirror effect of the windows and increases their normal light-transmission efficiency. The inner pane of each window is of chemically tempered, high-strength structural glass. The inner pane of the front windows has a seal (the docking window has two seals) between it and the window frame and is bolted to the frame through a metal retainer. The inner pane has the high-efficiency antireflective coating on its inner surface and a defogging coating on its outer surface.

All three windows are electrically heated to prevent fogging. The temperature of the windows is not monitored. Heater operation directly affects crew visibility; proper operation is, therefore, visually determined by the astronauts.

A window shade, with an antireflective coating on its outboard side, is provided for each window. Normally, the shade is rolled up at the window edge. A glareshield mounted between each front window and the control and flight display panels reduces window reflection of internal panel lighting.

EQUIPMENT STOWAGE

Underneath the control and display panels to the right of the LM Pilot's station is a compartmented cabinet. Equipment used during the mission is stowed in this cabinet. This equipment includes food, personal hygiene items, EVA waist tethers, a camera, camera lens filters, a spare light bulb, and a multipurpose special tool having a modified Allen-head. A similar cabinet to the left of the Commander's station contains a spare Environmental Control Subsystem (ECS) lithium

hydroxide canister, waste collection containers, a PLSS lithium hydroxide cartridge, and a PLSS condensate container.

MIDSECTION

From the flight stations, the astronauts have an 18-inch step up into the midsection, which is immediately aft of the crew compartment. Normally, the midsection is not manned; it is traversed by the astronauts upon entering and exiting the LM after docking. The midsection is 54 inches deep and approximately 5 feet high. The internal shape is elliptical, with a minor axis of approximately 56 inches. The midsection houses the ascent engine assembly, part of which protrudes up through the lower deck. ECS components, a water-dispensing fire extinguisher, a container for lunar samples, and life support and communications umbilicals are installed on the right side of the midsection. Along the left side, is the waste management system and an oxygen purge system. This side also contains stowage for food, lunar overshoes, a pilot's reference kit, and miscellaneous containers. Components of the Electrical Power Subsystem (EPS) and the Guidance, Navigation,

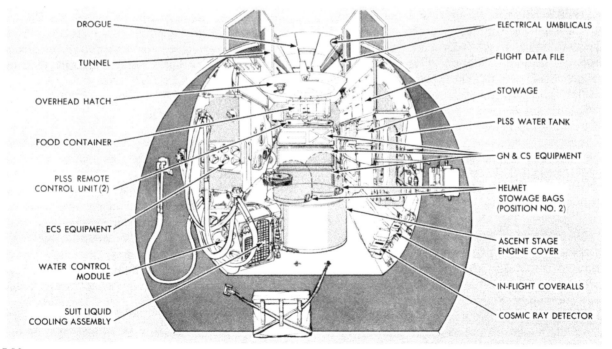

DROGUE
TUNNEL
OVERHEAD HATCH
FOOD CONTAINER
PLSS REMOTE CONTROL UNIT (2)
ECS EQUIPMENT
WATER CONTROL MODULE
SUIT LIQUID COOLING ASSEMBLY

ELECTRICAL UMBILICALS
FLIGHT DATA FILE
STOWAGE
PLSS WATER TANK
GN & CS EQUIPMENT
HELMET STOWAGE BAGS (POSITION NO. 2)
ASCENT STAGE ENGINE COVER
IN-FLIGHT COVERALLS
COSMIC RAY DETECTOR

R-5A

Midsection Interior

and Control Subsystem (GN&CS) are mounted on the aft bulkhead. A cylindrical cover protects the accessories section from the protruding part of the engine. The top of the cover is used as a rest position for one of the astronauts and as a platform for initially observing the lunar surface through the overhead (docking) hatch. Above the hatch is a docking tunnel; directly forward of the hatch, PLSS fittings are mounted. These fittings aid the astronauts in donning their PLSS units.

Construction of the midsection is similar to that of the crew compartment, but the midsection has a bulkhead at each end. The aft bulkhead supports the aft equipment bay structure. In addition to the lower deck to which the ascent engine is mounted, there are two others. One of these supports the overhead hatch and the lower end of the docking tunnel; the other, supports the upper end of the docking tunnel and absorbs some of the stresses imposed during docking. All decks are made of integrally stiffened machined aluminum alloy, or reinforced chemically milled web. The exterior structure forms a cradle around the midsection to absorb or transmit all stress loads applied to the ascent stage. Stress loads applied to beams on top of the crew compartment are transmitted through midsection beams, to the aft bulkhead and, in turn, to the interstage fittings. The external structure, along the sides of the midsection, supports propellant subsystem storage tanks and S-band steerable and VHF in-flight antennas. The aft midsection bulkhead supports propellant and ECS tanks, an aft equipment rack assembly and the Reaction Control System (RCS) two aft thrust clusters. A docking target, used for aligning the LM with the CSM during docking, is mounted on the upper left structure of the midsection exterior.

OVERHEAD HATCH

The overhead hatch, approximately 33 inches in diameter, is at the top centerline of the midsection. When the LM and CM are docked, the hatch permits transfer of astronauts and equipment. The astronauts pass through the hatch, head first. Handgrips in the docking tunnel immediately above the hatch aid in crew and equipment transfer. The hatch has an off-center latch that can be operated from either side of the hatch. The hatch

is opened inward by rotating the latch handle 90°. A preloaded elastomeric silicone compound seal is mounted in the hatch frame structure. When the latch is closed, a lip near the outer circumference of the hatch enters the seal, ensuring a pressure tight contact. Normal cabin pressurization forces the hatch into its seal. To open the hatch, the cabin must be depressurized by opening a cabin relief and dump valve, which is within the hatch structure. The valve can be operated with a handle on each side of the hatch.

DOCKING TUNNEL

The docking tunnel, immediately above the overhead hatch, provides a structural interface between the LM and the CM to permit transfer of equipment and astronauts without exposure to space environment. The tunnel is 32 inches in diameter and 16 inches long. A ring at the top of the tunnel is compatible with a docking ring on the CM. The CM docking ring has automatic clamping latches. The ring is concentric with the nominal centerline of thrust of the ascent and descent engines. The drogue, a portion of the docking mechanism, is secured below the ring to three mounts in the LM tunnel so that it can mate with the docking probe of the CM. When the CM and LM are docked, the rings are joined; this ensures structural continuity for transmitting midcourse correction and lunar orbit injection stresses throughout vehicle basic structure.

AFT EQUIPMENT BAY

The aft equipment bay is an unpressurized area formed by the aft midsection bulkhead and the equipment rack, which is cantilevered approximately 33 inches aft of the bulkhead. The equipment rack assembly has integral cold rails that transfer heat from electronic and electrical equipment (components of the GN&CS, EPS, and Communications Subsystems) mounted on the rack. The cold rails are mounted vertically in the rack structural frame. Water-glycol flows through the cold rails.

Two gaseous oxygen tanks and two gaseous helium tanks are secured to the truss members between the midsection aft bulkhead and the equipment rack. ECS and Main Propulsion Subsystem components that do not require a pressurized environment or acesss by the astronauts are mounted to the outboard side of the aft bulkhead. The equipment rack and the aft bulkhead support the aft RCS thrust clusters.

THERMAL AND MICROMETEOROID SHIELD

After the LM is removed from the spacecraft-Lunar Module adapter (SLA), it is exposed to micrometeoroids and solar radiation. To protect the LM astronauts and equipment from temperature extremes, active and passive thermal control is used. Active thermal control is provided by the ECS. Passive thermal control isolates the vehicle interior structure and equipment from its external environment to sustain acceptable temperature limits throughout the lunar mission. The entire ascent stage structure is enclosed within a thermal blanket and a micrometeoroid shield. Glass fiber

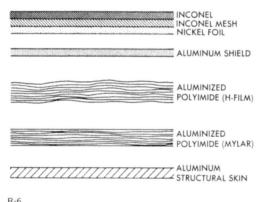

R-6

Typical Thermal Blanket and Micrometeoroid Shield

standoffs, of low thermal conductivity, hold the blanket away from the structural skin. Aluminum frames around the propellant tanks prevent contact between tanks and blanket. The thermal blanket consists of multiple-layered (at least 25 layers) of aluminized sheet (mylar or H-film). Each layer is only 0.00015 inch thick and is coated on one side with a microinch thickness of aluminum. To make an even more effective insulation, the polymide

sheets are hand crinkled before blanket fabrication. This crinkling provides a path for venting, and minimizes contact conductance between the layers. Structures with a high thermal conductivity, such as antenna supports and landing gear members, that pass through the thermal blanket also have thermal protection. Individual blanket layers are overlapped and sealed with a continuous strip of H-film tape. To join the multilayered sections, the

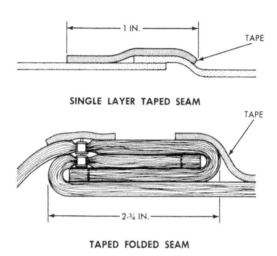

R-7

Thermal Blanket Joining Techniques

blanket edges are secured with grommet type fasteners, then the seam is folded and sealed with a continuous strip of tape. Mylar sheets are used predominantly in those areas where temperatures do not exceed 300° F. In areas where higher temperatures are sustained, additional layers of H-film are added to the mylar sheets. H-film can withstand temperatures up to 1000° F, but, because it is a heavier material, it is used only where absolutely necessary. Certain areas of the ascent stage are subjected to temperatures as high as 1800° F due to CSM and LM RCS plume impingement. These areas are thermally controlled by a sandwich material of thin nickel foil (0.0005 inch) interleaved with Inconel wire mesh and Inconel sheet. Finally, the highly reflective surfaces of the shades provided for the front and docking windows reduce heat absorption.

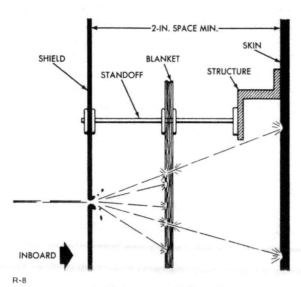

R-8

Typical Micrometeoroid Protection

The micrometeoroid shield, outboard of the thermal blanket, is a sheet of aluminum that varies in thickness from 0.004 to 0.008 inch, depending on micrometeoroid-penetration vulnerability. It is attached to the same standoffs as the thermal blankets. Various thermal control coatings are applied to the outer surface of the shield to provide an additional temperature boundary for vehicle insulation against space environment.

DESCENT STAGE

The descent stage is the unmanned portion of the LM; it represents approximately two-thirds of the weight of the LM at the earth-launch phase. This is because the descent engine is larger than the ascent engine and it requires a much larger propellant load. Additionally, its larger proportion of

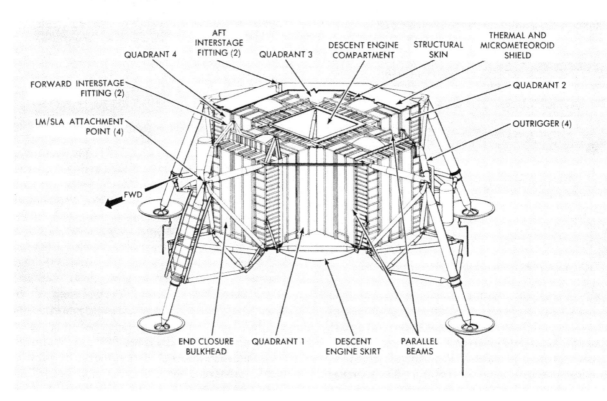

R-9A

Descent Stage Structure

weight results from necessity of the descent stage to:

(1) Support the entire ascent stage.
(2) Provide for attachment of the landing gear.
(3) Support the complete LM in the SLA.
(4) Provide structure to support the scientific and communications equipment to be used on the lunar surface.
(5) Act as the launching platform of the ascent stage.

The main structure of the descent stage consists of two pairs of parallel beams arranged in a cruciform, with a deck on the upper and lower surfaces, approximately 65 inches apart. The ends of the beams, approximately 81 inches from the center, are closed off by aluminum beams to provide five equally sized compartments: a center compartment, one forward and one aft of the center compartment, and one right and one left of the center compartment. A four-legged truss (outrigger) at the end of each pair of beams serves as a support for the LM in the SLA and as the attachment point for the upper end of the landing gear primary strut. Two of the four interstage fittings for attachment of the ascent stage are mounted on the forward compartment beams. The other two fittings are on the aft beam of the side compartments. The five compartments formed by the main beams house Main Propulsion Subsystem components. The center compartment houses the descent engine, which is supported by truss members and an engine gimbal ring. Descent engine fuel and oxidizer tanks are in the remaining compartments.

Struts between the ends of all main beams form triangular bays, or quadrants, to give the descent stage its octagon shape. The quadrants are designated 1 through 4, beginning at the left of the forward compartment and continuing counterclockwise (as viewed from the top) around the center. The quadrants house components from the various subsystems. In addition, the modularized equipment stowage assembly (MESA), in quad No. 4, and a pallet assembly is stowed in quad No. 3.

The MESA consists of television equipment, equipment for obtaining and stowing lunar samples, and PLSS components to be used by the astronauts during the lunar stay.

The quad No. 3 pallet assembly contains two pallets; a Lunar Roving Vehicle (LRV) pallet, and a pallet holding the Lunar Retro-Ranging Reflector. The LRV pallet contains a lunar geological exploration tool carrier, a lunar dust brush, a gnomen, a recording penetrometer, tongs, a trenching tool, collection bags, and other items needed during lunar exploration.

Four plume deflectors, which keep the plumes of the downward-firing RCS thrusters from impinging upon the descent stage, are truss-mounted to the descent stage.

THERMAL AND MICROMETEOROID SHIELD

The entire descent stage structure is enveloped in a thermal and micrometeoroid shield similar to that used on the ascent stage. Because the top deck and side panels of the descent stage are subjected to engine exhaust, these areas are extensively protected with a nickel inconel mesh sandwich outboard of the mylar and H-film blankets. A teflon-coated titanium blast shield that deflects the ascent engine exhaust out of and away from the descent engine compartment is secured to the upper side of the compartment, below the thermal blanket. Layers of H-film, joined to the blast deflector, act as an ablative membrane which protects the descent stage from ascent engine exhaust gases that are deflected outward, between the stages, during lift-off from the lunar surface. The engine compartment and the bottom of the descent stage are subjected to temperatures in excess of 1800°F when the descent engine is fired. A special base heat shield protects the descent stage structure and internal components. It consists of a titanium shield attached to descent stage structure. The heat shield supports a thermal blanket on each of its sides. The thermal blanket that faces the engine nozzle consists of multiple layers of nickel foil and glass wool and an outer layer of H-film. This blanket acts as a protective membrane to withstand engine exhaust gas back pressures at lunar touchdown and prevent heat, absorbed by the lunar surface during LM landing, from radiating back into the descent stage. Twenty-five layers of H-film make up the blanket on the other side of the titanium. A flange-like ring of columbium backed with a fibrous (Min-K) insulation is attached directly to the engine nozzle extension and joined to the base heat shield by an annular bellows of 25-layer H-film. This bellows arrangement permits descent engine gimbaling, but prevents engine heat from leaking into the engine compartment.

LANDING GEAR

The landing gear provides the impact attenuation required to land the LM on the lunar surface, prevents tipover of the LM on a lunar surface with a 6° general slope having 24-inch depressions or protuberances, and supports the LM during lunar stay and lunar launch. Landing impact is attenuated to load levels that preserves the LM structural integrity. At earth launch, the landing gear is retracted to reduce the overall size. It remains retracted until the docked CSM and LM attain lunar orbit and the astronauts have transferred to the LM. Before the LM is separated from the CSM, the Commander in the LM operates the landing gear deployment switch to extend the gear. At this time landing gear uplocks are explosively released, allowing springs in deployment mechanisms to extend the gear. Once extended, the landing gear is locked in place by downlock mechanisms.

The cantilever landing gear consists of four assemblies, each connected to an outrigger that extends from the ends of the structural parallel beams. The landing gear assemblies extend from the front, rear, and both sides of the descent stage. Each assembly consists of struts, trusses, a footpad, lock and deployment mechanisms, and, on all but the forward gear assembly, a lunar surface sensing probe. A ladder is affixed to the forward gear assembly.

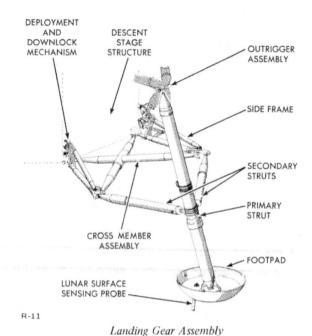

R-11

Landing Gear Assembly

The landing gear can withstand: (1) a 10-foot/second vertical velocity of the LM when the horizontal velocity is zero feet/second, (2) a 7-foot/second vertical velocity with a horizontal velocity not exceeding 4 feet/second, and (3) a vehicle attitude within 6° of the local horizontal when the rate of attitude change is 2°/second or less.

PRIMARY STRUT

The upper end of the primary strut is attached to the outboard end of the outrigger; the lower end has a ball joint for the footpad. The strut is of the piston-cylinder type; it absorbs the compression load of the lunar landing and supports the LM on the lunar surface. Compression loads are attenuated by a crushable aluminum-honeycomb cartridge in each strut. Maximum compression length of the primary strut is 32 inches. The aluminum honeycomb has the shock-absorbing capability of accepting one lunar landing. This may include one or two bounces of the LM, but after the full weight of the LM is on the gear, the shock-absorbing medium is expended. Use of compressible honeycomb cartridges eliminated the need for thick-walled, heavyweight, pneudraulic-type struts.

GRUMMAN

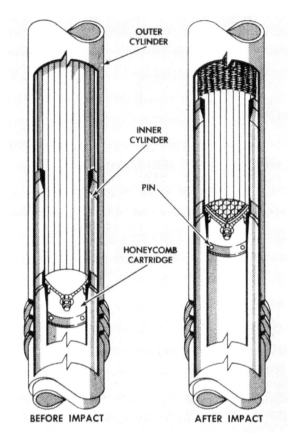

OUTER CYLINDER

INNER CYLINDER

PIN

HONEYCOMB CARTRIDGE

BEFORE IMPACT   AFTER IMPACT

R-12

Landing Gear Primary Strut

The footpad, attached to the strut by a ball-socket fitting, is aluminum-honeycomb; its diameter is 37 inches. This large diameter ensures minimal penetration of the LM on low load-bearing-strength lunar surface. During earth launch, four restraining straps hold the pads in a fixed position on the strut. The straps shear or bend on pad contact with the lunar surface, permitting the pad to conform to surface irregularities.

LUNAR SURFACE SENSING PROBE

The lunar surface sensing probe attached to each landing gear footpad, except the forward one, is an electromechanical device. The probes are retained in the stowed position, against the primary strut, until landing gear deployment. During deployment, mechanical interlocks are released permitting spring energy to extend the probes so that the probe head is approximately 5 feet below the footpad. When any probe touches the lunar surface,

pressure on the probe head will complete the circuit that advises the astronauts to shut down the descent engine. This shutdown point which determines LM velocity at impact, is a tradeoff between landing gear design weight and the thermal and thrust reactions caused by the descent engine operating near the lunar surface. Each probe has indicator plates attached to it, which, when aligned, indicate that the probes are fully extended.

SECONDARY STRUTS

Each landing gear assembly has two secondary struts. The outboard end of each strut is attached to the primary strut; the inboard ends are attached to a deployment truss assembly. Each strut is a piston-cylinder-type device that contains compressible aluminum honeycomb capable of absorbing compression and tension loads. The design and the location of the secondary struts in relation to the primary strut enables the LM to land on an unsymmetrical surface or to land when the LM is moving laterally over the lunar surface.

UPLOCK ASSEMBLY

One uplock assembly is attached to each landing gear assembly. It consists of a fixed link (strap) and two end detonator cartridges in a single case. The fixed link, attached between the primary strut and the descent stage structure, holds the landing gear in its retracted position. When the Commander operates the landing gear deployment switch, it activates an electrical circuit which explosively severs the fixed link to permit the deployment mechanism to extend the landing gear. When detonated, either end cartridge has sufficient energy to sever the fixed link.

DEPLOYMENT AND DOWNLOCK MECHANISM

The deployment portion of the deployment and downlock mechanism consists of a truss assembly, two clock-type deployment springs, and connecting linkage. The truss, connecting the secondary struts and descent stage structure, comprises two side frame assemblies separated by a crossmember. The deployment springs are attached, indirectly, to the side frame assemblies through connecting

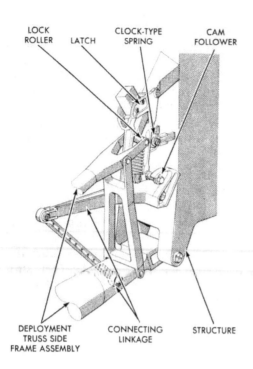

R-13

Deployment and Downlock Mechanism

linkage. The downlock portion of the mechanism consists of a spring-loaded lock and a cam follower. The follower rides on a cam attached to the deployment portion of the mechanism. When the fixed link of the uplock assembly is severed, the deployment springs pull the connecting linkage and, indirectly, the deployment truss. This action drives the landing gear from the stowed to the fully deployed position. At full gear deployment, the cam follower reaches a point that permits the spring-loaded lock to snap over a roller on the truss assembly. The lock cannot be opened. A landing gear deployment talkback advises the astronauts that the landing gear is fully deployed.

LADDER

The ladder affixed to the primary strut of the forward landing gear assembly has nine rungs between two railings. The rungs are spaced nine inches apart; the railings have approximately 20 inches between centers. The top of the ladder is approximately 18 inches below the forward end of the platform on the outrigger; the ladder extends down to within 30 inches of the footpad. This allows the primary strut to telescope when the LM impacts on the lunar surface.

PLATFORM

An external platform, approximately 32 inches wide and 45 inches long is mounted over the forward landing gear outrigger. The platform is just below the forward hatch. The upper surface is corrugated to facilitate hand and foot holds. The platform, in conjunction with the ladder below it provides the astronauts with a means of access between the vehicle and the lunar surface and between the LM interior and free space for EVA.

INTERFACES

At earth launch, the LM is housed within the SLA, which has an upper and a lower section. The upper section has deployable panels, which are jettisoned; the lower section has fixed panels. The upper panels are deployed and jettisoned when the CSM is separated from the SLA. During this separation phase, an explosive charge separates an umbilical line that connectes the LM, SLA, and launch umbilical tower. Before earth launch, this umbilical enables monitoring, purging, and control of the LM environment.

After transposition, the CSM docks with the LM. A ring at the top of the ascent stage docking tunnel provides a structural interface for joining the CSM to the LM. The ring is compatible with a clamping mechanism in the CSM docking ring. A drogue, which mates with the CSM docking probe, is installed in the docking tunnel, just below the ring. The probe provides initial vehicle soft docking and attenuates impact imposed by contact of the CSM and LM. After the CSM probe and drogue have joined, latches around the periphery of the CSM docking ring engage to effect full structural continuity and a pressure-tight seal between the vehicles. After docking has been completed, the astronauts connect electrical umbilicals in the CSM and the LM. These umbilicals provide electrical power to the LM, for separation from the SLA.

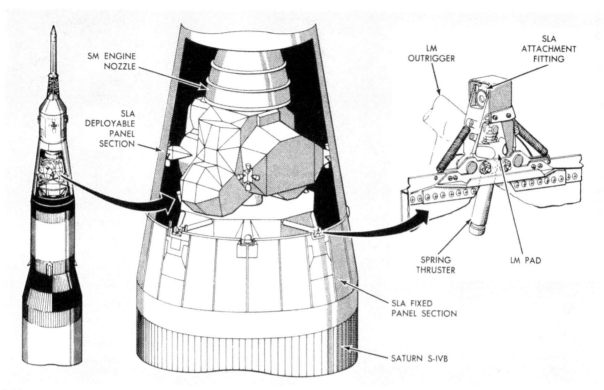

R-14

LM Interface with SLA

Pads at the apex of the descent stage outriggers mate with attachment fittings in the SLA. Tiedown tension straps, which are explosively released hold the pads against the LSA attachment fittings. When the CSM and the LM have docked, an astronaut in the CM initiates severence of the tiedown straps. After the straps are severed, preloaded spring thrusters provide positive separation of the LM from the SLA.

EXPLOSIVE DEVICES

The explosive devices are electro chemical devices, which are operated by the astronauts to perform the following functions:

Propellant tank pressurization, so that the ascent engine, descent engine, and Reaction Control Subsystem can be operated

Ascent and descent stage separation to allow the ascent stage to take off from the lunar surface, or for a mission abort

Descent propellant tank venting after landing

Landing gear deployment

The LM has two types of explosive devices. Generally, these devices consist of detonator cartridges, containing high-explosive charges of high yield; and pressure cartridges, containing propellant charges of relatively low yield. An electrical signal, originated by the Commander through control switches, triggers an initiator, which fires the cartridges.

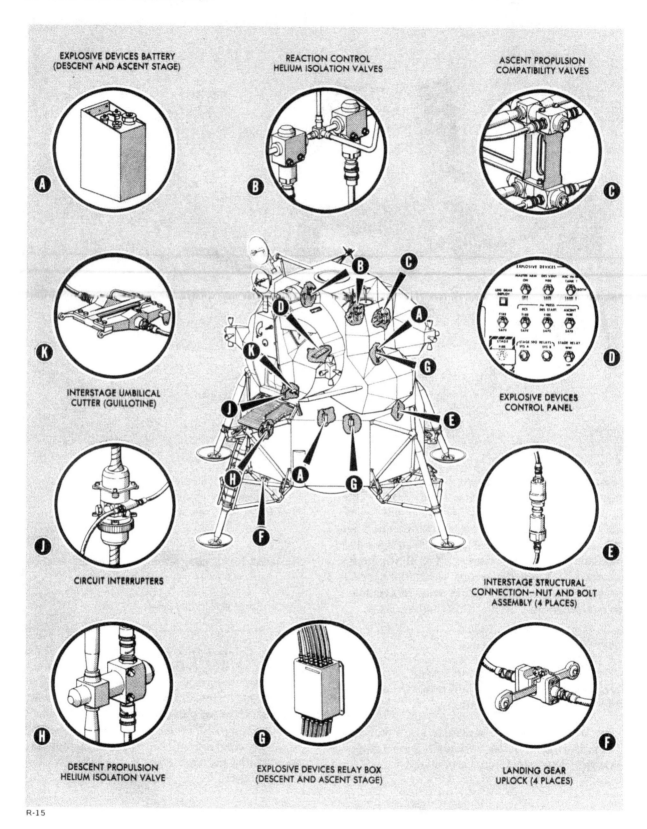

Major Explosive Devices Equipment Location

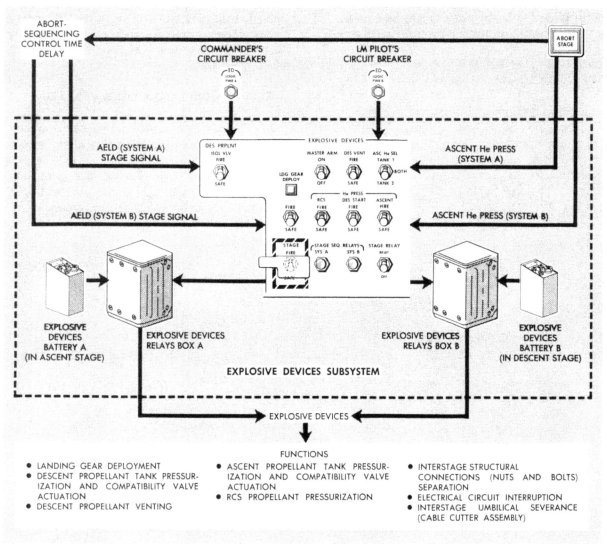

Diagram of Explosive Devices

R-16

LANDING GEAR DEPLOYMENT

Each of the four landing gear assemblies is restrained in the stowed position by an uplock assembly that contains two detonator cartridges. While the LM is docked with the CM in lunar orbit, the LM Commander fires both detonator cartridges in each uplock assembly to deploy the landing gear. When all four landing gear assemblies have been deployed, a landing gear deployment indicator flag (talkback) on the control panel turns gray.

REACTION CONTROL SUBSYSTEM PROPELLANT TANK PRESSURIZATION

The Reaction Control Subsystem fuel and oxidizer tanks are pressurized immediately before landing gear deployment. The Commander fires two cartridges, which open dual, parallel helium isolation valves, to pressurize the tanks. The subsystem can then be operated to separate the LM from the CM.

DESCENT PROPELLANT TANK PRESSURIZATION

After the LM and CM separate, the astronauts start the descent engine. However, before initiating the start, the descent propellant tanks must be pressurized. The Commander fires the compatibility valve cartridges, explosively opening the valves. He then fires cartridges to explosively open the ambient helium isolation valve. After the descent engine is started, the cryogenic helium flows freely to the descent engine fuel and oxidizer tanks, pressurizing them.

DESCENT PROPELLANT TANK VENTING

After lunar landing, the Commander simultaneously opens two explosive vent valves to accomplish planned depressurization of the descent propellant tanks. This protects the astronauts, when outside the LM, against untimely venting of the tanks through the relief valve assemblies.

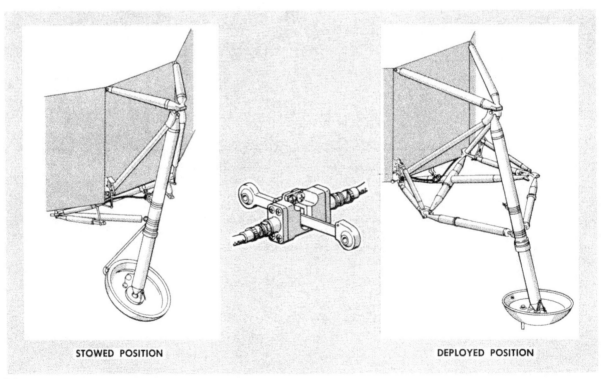

STOWED POSITION DEPLOYED POSITION

R-17

Landing Gear Deployment

ASCENT PROPELLANT TANK
PRESSURIZATION

After lunar stay, the astronauts take off from the lunar surface in the ascent stage. This requires pressurization of the ascent propellant tanks shortly before initial start of the ascent engine. To accomplish pressurization, the Commander fires explosive valve cartridges, which simultaneously open helium isolation valves and fuel and oxidizer compatibility valves. This permits helium to pressurize the ascent fuel and oxidizer tanks.

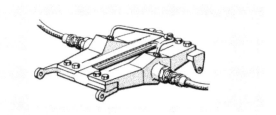

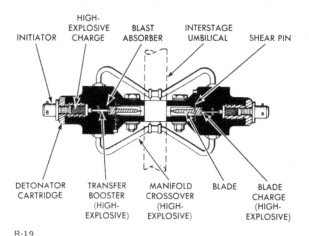

R-19

Explosive Guillotine

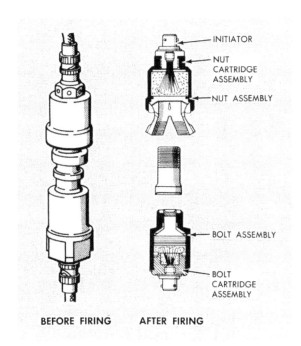

BEFORE FIRING   AFTER FIRING

R-18

Explosive Nuts and Bolts

STAGE SEPARATION

The ascent and descent stages are separated immediately before lunar lift-off or in the event of mission abort. The Commander sets control switches to initiate a controlled sequence of stage separation. First, all signal and electrical power between the two stages is terminated by explosive circuit interrupters. Next, explosive nuts and bolts joining the stages are ignited. Finally, an explosive guillotine (cable cutter assembly) automatically severs all wires, cables, and water lines connected between the stages. Stage separation completed, operation of its engine can propel the ascent stage into lunar orbit for rendezvous with the CM.

CREW PERSONAL EQUIPMENT

Crew personal equipment includes a variety of mission-oriented equipment required for life support and astronaut safety and accessories related to successful completion of the mission.

These equipments range from astronaut space suits and docking aids to personal items stored throughout the cabin. The Modularized Equipment Stowage Assembly (MESA), Apollo lunar scientific experiments payload (ALSEP), Quad 3 pallet assembly, and the Lunar Roving Vehicle (LRV) are stored in the descent stage.

This equipment is used for sample and data collecting and scientific experimenting. The resultant data will be used to derive information on the atmosphere and distance between earth and moon.

The portable life support system (PLSS) interfaces with the Environmental Control Subsystem (ECS) for refills of oxygen and water. The pressure garment assembly (PGA) interfaces with the ECS for conditioned oxygen, through oxygen umbilicals, and with the Communications and Instrumentation Subsystems for communications and bioinstrumentation, through the electrical umbilical.

EXTRAVEHICULAR MOBILITY UNIT

The extravehicular mobility unit (EMU) provides life support in a pressurized or unpressurized cabin, and up to 7 hours of extravehicular life support (depending on astronaut's metabolic rate).

In its extravehicular configuration, the EMU is a closed-circuit pressure vessel that envelops the astronaut. The environment inside the pressure vessel consists of 100% oxygen at a nominal pressure of 3.75 psia. The oxygen is provided at a flow rate of 6 cfm. The extravehicular life support equipment configuration includes the following:

    Liquid cooling garment (LCG)
    Pressure garment assembly (PGA)
    Integrated thermal micrometeoroid garment (ITMG)
    Portable life support system (PLSS)
    Oxygen purge system (OPS)
    Communications carrier
    EMU waste management system
    EMU maintenance kit
    PLSS remote control unit
    Lunar extravehicular visor assembly (LEVA)
    Biomedical belt

LIQUID COOLING GARMENT

The liquid cooling garment (LCG) is worn by the astronauts while in the LM and during all extravehicular activity. It cools the astronaut's body during extravehicular activity by absorbing body heat and transferring excessive heat to the sublimator in the PLSS. The LCG is a one-piece, long-sleeved, integrated-stocking undergarment of netting material. It consists of an inner liner of Beta cloth, to facilitate donning, and an outer layer of Beta cloth into which a network of Tygon tubing is woven. The tubing does not pass through the stocking area. A double connector for incoming and outgoing water is located on the front of the garment. Cooled water, supplied from the PLSS, is pumped through the tubing. Pockets for bioinstrumentation signal conditioners are located around the waist. A zipper that runs up the front is used for donning and doffing the LCG; an opening at the crotch is used for urinating. Dosimeter pockets and snaps for attaching a biomedical belt are part of the LCG.

PRESSURE GARMENT ASSEMBLY

The pressure garment assembly (PGA) is the basic pressure vessel of the EMU. It provides a mobile life-support chamber if cabin pressure is lost due to leaks or puncture of the vehicle. The PGA consists of a helmet, torso and limb suit, intravehicular gloves, and various controls and instrumentation to provide the crewman with a controlled environment. The PGA is designed to be worn for 115 hours; in an emergency, at a regulated pressure of 3.75± 0.25 psig, in conjunction with the LCG.

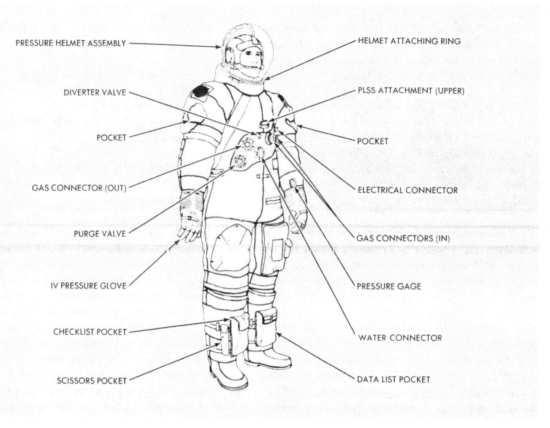

PRESSURE HELMET ASSEMBLY

DIVERTER VALVE

POCKET

GAS CONNECTOR (OUT)

PURGE VALVE

IV PRESSURE GLOVE

CHECKLIST POCKET

SCISSORS POCKET

HELMET ATTACHING RING

PLSS ATTACHMENT (UPPER)

POCKET

ELECTRICAL CONNECTOR

GAS CONNECTORS (IN)

PRESSURE GAGE

WATER CONNECTOR

DATA LIST POCKET

R-20A

Pressure Garment Assembly

The torso and limb suit is a flexible pressure garment that encompasses the entire body, except the head and hands. It has four gas connectors, a PGA multiple water receptacle, a PGA electrical connector, and a PGA urine transfer connector for the PLSS/PGA and ECS/PGA interface. The PGA connectors have positive locking devices and can be connected and disconnected without assistance. The gas connectors comprise an oxygen inlet and outlet connector, on each side of the suit front torso. Each oxygen inlet connector has an integral ventilation diverter valve. The PGA multiple water receptacle, mounted on the suit torso, serves as the interface between the LCG multiple water connector and PLSS multiple water connector. A protective external cover provides PGA pressure integrity when the LCG multiple water connector is removed from the PGA water receptacle. The PGA electrical connector, provides a communications, instrumentation, and power interface to

the PGA. The PGA urine transfer connector on the suit right leg is used to transfer urine from the urine collection transfer assembly (UCTA) to the waste management system.

The urine transfer connector, permits dumping the urine collection bag without depressurizing the PGA. A pressure relief valve on the right-leg thigh vents the suit in the event of overpressurization. If the valve does not open, it can be manually overridden. A pressure gage on the left sleeve indicates suit pressure.

The helmet is a Lexan (polycarbonate) shell with a bubble-type visor, a vent-pad assembly, and a helmet attaching ring. The vent-pad assembly permits a constant flow of oxygen over the inner front surface of the helmet. The astronaut can turn his head within the helmet neck-ring area. The helmet does not turn independently of the torso and limb suit. The helmet has provisions on each side for mounting a lunar extravehicular visor assembly (LEVA). When the LM is unoccupied, the helmet protective bags are stowed on the cabin floor at the crew flight stations. Each bag has a hollow-shell plastic base with a circular channel for the helmet and the LEVA, two recessed holes for glove connector rings, and a slot for the EMU maintenance kit. The bag is made of Beta cloth, with a circumferential zipper; it folds toward the plastic base when empty.

The intravehicular gloves are worn during operations in the LM cabin. The gloves are secured to the wrist rings of the torso and limb suit with a slide lock; they rotate by means of a ball-bearing race. Freedom of rotation, along with convoluted bladders at the wrists and adjustable anti-ballooning restraints on the knuckle areas, permits manual operations while wearing the gloves.

All PGA controls are accessible to the crewman during intravehicular and extravehicular operations. The PGA controls comprise two ventilation diverter valves, a pressure relief valve with manual override, and a manual purge valve. For intravehicular operations, the ventilation diverter valves are open, dividing the PGA inlet oxygen flow equally between the torso and helmet of the PGA. During extravehicular operation, the ventilation diverter valves are closed and the entire oxygen flow enters the helmet. The pressure relief valve accommodates flow from a failed-open primary oxygen pressure regulator. If the pressure relief valve fails open, it may be manually closed. The purge valve interfaces with the PGA through the PGA oxygen outlet connector. Manual operation of this valve initiates an 8 pound/hour purge flow, providing CO_2 washout and minimum cooling during contingency or emergency operations.

A pressure transducer on the right cuff indicates pressure within the PGA. Biomedical instrumentation comprises an EKG (heart) sensor, ZPN (respiration rate) sensor, dc-to-dc converter, and wiring harness. A personal radiation dosimeter (active) is attached to the integrated thermal micrometeoroid garment for continuous accumulative radiation readout. A chronograph wristwatch (elapsed-time indicator) is readily accessible to the crewman for monitoring.

COMMUNICATIONS CARRIER

The communications carrier (cap) is a polyurethane-foam headpiece with two independent earphones and microphones, which are connected to the suit 21-pin communications electrical connector. The communications carrier is worn with or without the helmet during intravehicular operations. It is worn with the helmet during extravehicular operations.

INTEGRATED THERMAL MICROMETEOROID GARMENT

The ITMG, worn over the PGA, protects the astronaut from harmful radiation, heat transfer, and micrometeoroid activity. It is a one-piece, form-fitting, multilayered garment that is laced over the PGA and remains with it. The LEVA, gloves, and boots are donned separately. From the outer layer in, the ITMG is made of a protective cover, a micrometeoroid-shielding layer, a thermal-barrier blanket (multiple layers of aluminized mylar), and a protective liner. A zipper on the ITMG permits connecting or disconnecting umbilical hoses. For extravehicular activity, the PGA gloves are replaced with the extravehicular gloves. The extravehicular gloves are made of the same material as the ITMG to permit handling intensely hot or cold objects outside the cabin and for protection against lunar temperatures. The extravehicular boots (lunar overshoes) are worn over the PGA boots for extravehicular activity. They are made of the same material as the ITMG. The soles have additional insulation for protection against intense temperatures.

The LEVA, which fits over the clamps around the base of the helmet; provides added protection against solar heat, space particles, solar glare, ultraviolet rays, and accidental damage to the helmet. The LEVA is comprised of a plastic shell, cover, hinge assemblies, three eyeshades, and two visors (protective and sun visors). The protective visor

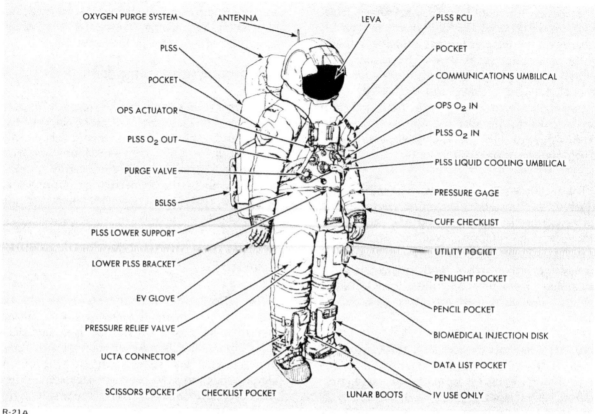

OXYGEN PURGE SYSTEM — ANTENNA — LEVA — PLSS RCU

PLSS — POCKET

POCKET — COMMUNICATIONS UMBILICAL

OPS ACTUATOR — OPS O$_2$ IN

PLSS O$_2$ OUT — PLSS O$_2$ IN

PURGE VALVE — PLSS LIQUID COOLING UMBILICAL

BSLSS — PRESSURE GAGE

PLSS LOWER SUPPORT — CUFF CHECKLIST

LOWER PLSS BRACKET — UTILITY POCKET

EV GLOVE — PENLIGHT POCKET

PRESSURE RELIEF VALVE — PENCIL POCKET

UCTA CONNECTOR — BIOMEDICAL INJECTION DISK

SCISSORS POCKET — CHECKLIST POCKET — LUNAR BOOTS — DATA LIST POCKET — IV USE ONLY

R-21A

Integrated Thermal Micrometeoroid Garment

provides impact, infrared, and ultra-violet ray protection. The sun visor has a gold coating which provides protection against light and reduces heat gain within the helmet. The eyeshades, two located on each side and one in the center, reduces low-angle solar glare by preventing light penetration at the sides and overhead viewing area. When the LM is occupied, the LEVA's are stowed in helmet stowage bags and secured on the ascent engine cover.

PORTABLE LIFE SUPPORT SYSTEM

The PLSS is a self-contained, self-powered, rechargeable environmental control system. In the extravehicular configuration of the EMU, the PLSS is worn on the astronaut's back. The PLSS supplies pressurized oxygen to the PGA, cleans and cools the expired gas, circulates cool liquid in the LCG through the liquid transport loop, transmits astronaut biomedical data, and functions as a duel VHF transceiver for communication.

The PLSS has a contoured fiberglass shell to fit the back, and a thermal micrometeoroid protective cover. It has three control valves, and, on a separate remote control unit, two control switches, a volume control, and a five-position switch for the dual VHF transceiver. The remote control unit is set on the chest.

The PLSS attaches to the astronaut's back, over the ITMG; it is connected by a shoulder harness assembly. When not in use, it is stowed on the floor or in the left-hand midsection. To don the PLSS, it is first hooked to the overhead attachments in the left-hand midsection ceiling. The astronaut backs against the pack, makes PGA and harness connections, and unhooks the PLSS straps from the overhead attachment.

The PLSS can operate for 7 hours, depending upon the astronauts metabolic rate, before oxygen and feedwater must be replenished and the battery

GRUMMAN

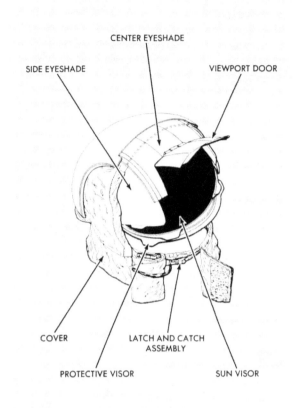

Lunar Extravehicular Visor Assembly

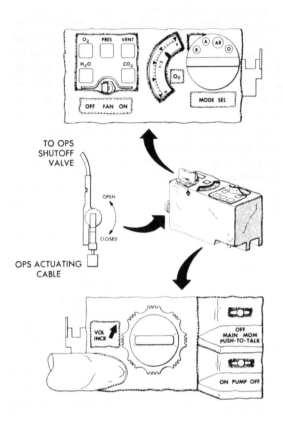

PLSS Remote Control Unit

replaced. The basic systems and loops of the PLSS are primary oxygen subsystem, oxygen ventilation loop, feedwater loop, liquid transport loop, and electrical system.

The space suit communicator (SSC) in the PLSS provides primary and secondary duplex voice communication and physiological and environmental telemetry. All EMU data and voice must be relayed through the LM and CM and transmitted to MSFN via S-band. The VHF antenna is permanently mounted on the oxygen purge system (OPS). Two tone generators in the SSC generate audible 3- and 1.5-kHz warning tones to the communications cap receivers. The generators are automatically turned on by high oxygen flow, low vent flow, or low PGA pressure. Both tones are readily distinguishable.

PLSS REMOTE CONTROL UNIT

The PLSS remote control unit is a chest-mounted instrumentation and control unit. It has a fan switch, pump switch, SSC mode selector switch, volume control, PLSS oxygen quantity indicator, five status indicators, and an interface for the OPS actuator.

OXYGEN PURGE SYSTEM

The OPS is a self-contained, independently powered, high-pressure, nonrechargeable emergency oxygen system that provides 30 minutes of regulated purge flow. The OPS consists of two interconnected spherical high-pressure oxygen bottles, an automatic temperature control module, an oxygen pressure regulator assembly, a

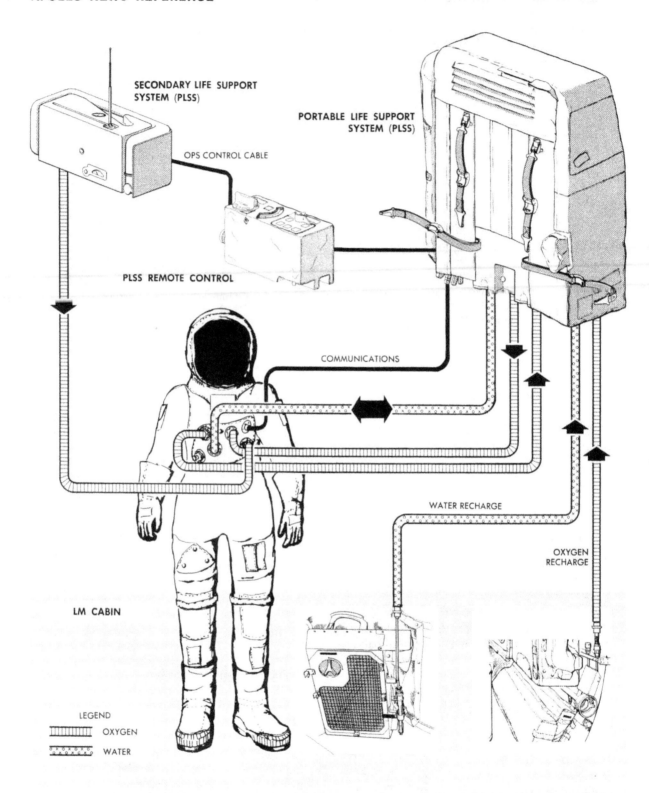

SECONDARY LIFE SUPPORT SYSTEM (PLSS)

OPS CONTROL CABLE

PORTABLE LIFE SUPPORT SYSTEM (PLSS)

PLSS REMOTE CONTROL

COMMUNICATIONS

WATER RECHARGE

OXYGEN RECHARGE

LM CABIN

LEGEND

OXYGEN

WATER

R-24A

Diagram of the Portable Life Support System

battery, an oxygen connector, and checkout instrumentation. In the normal extravehicular configuration, the OPS is mounted on top of the PLSS and is used with PLSS systems during emergency operations. In the contingency extravehicular configuration, the OPS is attached to the PGA front lower torso and functions independently of the PLSS. The OPS has no communications capability, but provides a hard mount for the SSC VHF antenna. Two OPS's are stowed in the LM.

UMBILICAL ASSEMBLY

The umbilical assembly consists of hoses and connectors for securing the PGA to the ECS, Communications Subsystem (CS), and Instrumentation Subsystem (IS). Separate oxygen and electrical umbilicals connect to each astronaut.

The oxygen umbilical consists of Flourel hoses (1.25-inch inside diameter) with wire reinforcement. The connectors are of the quick-disconnect type, with a 1.24-inch, 90° elbow at the PGA end. Each assembly is made up of two hoses and a dual-passage connector at the ECS end and two separate hoses (supply and exhaust) at the PGA end. When not connected to the PGA, the ECS connector end remains attached and the hoses stowed.

The electrical umbilical carries voice communications and biomedical data, and electrical power for warning-tone impulses.

CREW LIFE SUPPORT

The crew life support equipment includes food and water, a waste management system, personal hygiene items, and pills for in-flight emergencies. A potable-water unit and food packages contain sufficient life-sustaining supply for completion of the LM mission.

CREW WATER SYSTEM

The water dispenser assembly consists of a mounting bracket, a coiled hose, and a trigger-actuated water dispenser. The hose and dispenser extend approximately 72 inches to dispense water from the ECS water feed control assembly. The ECS water feed control valve is opened to permit water flow. The dispenser assembly supplies water at +50° to +90° for drinking or food preparation and fire extinguishing. The water for drinking and food preparation is filtered through a bacteria filter. The water dispenser is inserted directly into the mouth for drinking. Pressing the trigger-type control supplies a thin stream of water for drinking and food preparation. For firefighting, a valve on the dispenser is opened. The valve provides a greater volume of water than that required for drinking and food preparation.

FOOD PREPARATION AND CONSUMPTION

The astronaut's food supply (approximately 3,500 calories per man per day) includes liquids and solids with adequate nutritional value and low waste content. Food packages are stowed in the LM midsection, on the shelf above PLSS No. 1 and the right-hand stowage compartment and the MESA.

The food is vacuum packed in plastic bags that have one-way poppet valves into which the water dispenser can be inserted. Another valve allows food passage for eating. The food bags are packaged in aluminum-foil-backed plastic bags for stowage and are color coded: red (breakfast), white (lunch), and blue (snacks).

Food preparation involves reconstituting the food with water. The food bag poppet-valve cover is cut with scissors and pushed over the water dispenser nozzle after its protective cover is removed. Pressing the water dispenser trigger releases water. The desired consistency of the food determines the quantity of water added. After withdrawing the water dispenser nozzle, the protective cover is replaced and the dispenser returned to its stowage position. The food bag is kneaded for approximately 3 minutes, after which the food is considered reconstituted. After cutting off the neck of the food bag, food can be squeezed into the mouth through the food-passage valve. A germicide tablet, attached to the outside of the food bag, is inserted into the bag after food consumption, to prevent fermentation and gas formation. The bag is rolled to its smallest size, banded, and placed in the waste disposal compartment.

EMU WASTE MANAGEMENT SYSTEM

The EMU waste management system provides for the disposal of body waste through use of a fecal containment system and a urine collection and transfer assembly, and for neutralizing odors. Personal hygiene items are stowed in the left-hand stowage compartment.

Waste fluids are transferred to a waste fluid collector assembly by a controlled difference in pressures between the PGA and cabin (ambient). The primary waste fluid collector consists of a long transfer hose, control valve, short transfer hose, and a 8,900-cc multilaminate bag. The long transfer hose is stowed on a connector plate when not in use. To empty his in-suit urine container, the astronaut attaches the hose to the PGA quick-disconnect, which has a visual flow indicator. Rotating the handle of the spring-loaded waste control valve controls passage of urine to the assembly. The 8,900-cc bag is in the PLSS LiOH storage unit, the short transfer hose is connected between the waste control valve and the bag.

With cabin pressure normal (4.8 psia), the long transfer hose is removed from the connector stowage plate and attached to the PGA male disconnect. The PGA is overpressurized by 0.8 ± 0.2 psia and the waste control valve is opened. Urine flows from the PGA to the collector assembly at a rate of approximately 200 cc per minute. When bubbles appear in flow indicator, the valve indicator is released and allowed to close.

A secondary waste fluid collector system provides 900-cc waste fluid containers, which attach directly to the PGA. Urine is transferred directly from the PGA, through the connectors, to the bags. These bags can then be emptied into the 8,900-cc collector assembly.

FECAL DEVICE

The fecal containment system consists of an outer fecal/emesis bag (one layer of Aclar) and a smaller inner bag. The inner bag has waxed tissue on its inner surface. Polyethylene-backed toilet tissue and a disinfectant package are stored in the inner bag.

To use, the astronaut removes the inner bag from the outer bag. After unfastening the PGA and removing undergarments, the waxed tissue is peeled off the bag's inner surface and the bag is placed securely on the buttocks. After use, the used toilet tissue is deposited in the used bag and the disinfectant package is pinched and broken inside the bag. The bag is then closed, kneaded, and inserted in the outer bag. The wax paper is removed from the adhesive on the fecal/emesis bag and the bag is sealed then placed in the waste disposal compartment.

PERSONAL HYGIENE ITEMS

Personal hygiene items consist of wet and dry cleaning cloths, chemically treated and sealed in plastic covers. The cloths measure 4 by 4 inches and are folded into 2-inch squares. They are stored in the food package container.

MEDICAL EQUIPMENT

The medical equipment consists of biomedical sensors, personal radiation dosimeters, and emergency medical equipment.

Biomedical sensors gather physiological data for telemetry. Impedance pneumographs continuously record heart beat (EKG) and respiration rate. Each assembly (one for each astronaut) has four electrodes which contain electrolyte paste; they are attached with tape to the astronaut's body.

Six personal radiation dosimeters are provided for each astronaut. They contain thermoluminescent powder, nuclear emulsions, and film that is sensitive to beta, gamma, and neutron radiation. They are placed on the forehead or right temple, chest, wrist, thigh, and ankle to detect radiation to eyes, bone marrow, and skin. Serious, perhaps critical, damage results if radiation dosage exceeds a predetermined level. For quick, easy reference each astronaut has a dosimeter mounted on his EMU.

The emergency medical equipment consists of a kit of six capsules: four are pain killers (Darvon) and two are pep pills (Dexedrine). The kit is attached to the interior of the flight data file, readily accessible to both astronauts.

CREW SUPPORT AND RESTRAINT EQUIPMENT

The crew support and restraint equipment includes armrests, handholds (grips), Velcro on the floor to interface with the boots, and a restraint assembly operated by a rope-and-pulley arrangement that secures the astronauts in an upright position under zero-g conditions.

The armrests, at each astronaut position, provide stability for operation of the thrust/translation controller assembly and the attitude controller assembly, and restrain the astronaut laterally. They are adjustable (four positions) to accommodate the astronaut; they also have stowed (fully up) and docking (fully down) positions. The armrests, held in position by spring-loaded detents, can be moved from the stowed position by grasping them and applying downward force. Other positions are selected by pressing latch buttons on the armrest forward area. Shock attenuators are built into the armrests for protection against positive-g forces (lunar landing). The maximum energy absorption of the armrest assembly is a 300-pound force, which will cause a 4-inch armrest deflection.

The handholds, at each astronaut station and at various locations around the cabin, provide support for the upper torso when activity involves turning, reaching, or bending; they attenuate movement in any direction. The forward

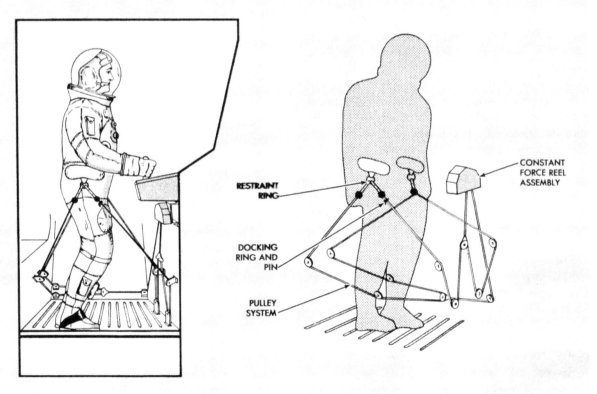

R-25

Restraint Equipment

panel handholds are single upright, peg-type, metal grips. They are fitted into the forward bulkhead, directly ahead of the astronauts, and can be grasped with the left or right hand.

The restraint assembly consists of cables, restraint rings, and a constant-force reel system. The cables attach to D-rings on the PGA sides, waist high. The constant-force reel provides a downward force of approximately 30 pounds, it is locked during landing or docking operations. When the constant-force reel is locked, the cables are free to reel in. A ratchet stop prevents paying out of the cables and thus provides zero-g restraint. During docking maneuvers, the Commander uses pin adjustments to enable him to use the crewmen optical alignment sight (COAS) at the overhead (docking) window.

DOCKING AIDS AND TUNNEL HARDWARE

Docking operations require special equipment and tunnel hardware to effect linkup of the LM with the CSM. Docking equipment includes the crewman's optical alignment sight (COAS) and a docking target. A drogue assembly, probe assembly; the CSM forward hatch, and hardware inside the LM tunnel enable completion of the docking maneuver.

The COAS provides the Commander with gross range cues and closing rate cues during the docking maneuver. The closing operation, from 150 feet to contact, is an ocular, kinesthetic coordination that requires control with minimal use of fuel and time. The COAS provides the Commander with a fixed line-of-sight attitude reference image, which appears to be the same distance away as the target.

The COAS is a collimating instrument. It weighs approximately 1.5 pounds, is 8 inches long, and operates from a 28-volt d-c power source. The COAS consists of a lamp with an intensity control, a reticle, a barrel-shaped housing and mounting track, and a combiner and power receptacle. The reticle has vertical and horizontal 10° gradations in a 10° segment of the circular combiner glass, on an elevation scale

(right side) of –10° to +31.5°. The COAS is capped and secured to its mount above the left window (position No. 1).

To use the COAS, it is moved from position No. 1 to its mount on the overhead docking window frame (position No. 2) and the panel switch is set from OFF to OVHD. The intensity control is turned clockwise until the reticle appears on the combiner glass; it is adjusted for required brightness.

The docking target permits docking to be accomplished on a three-dimensional alignment basis. The target consists of an inner circle and a standoff cross of black with self-illuminating disks within an outer circumference of white. The target-base diameter is 17.68 inches. The standoff cross is centered 15 inches higher than the base and, as seen at the intercept, is parallel to the X-axis and perpendicular to the Y-axis and the Z-axis.

The drogue assembly consists of a conical structure mounted within the LM docking tunnel. It is secured at three points on the periphery of the tunnel, below the LM docking ring. The LM docking ring is part of the LM midsection outer structure, concentric with the X-axis. The drogue assembly can be removed from the CSM end or LM end of the tunnel.

Basically, the assembly is a three-section aluminum cone secured with mounting lugs to the LM tunnel ring structure. A lock and release mechanism on the probe, controls capture of the CSM probe at CSM-LM contact. Handles are provided to release the drogue from its tunnel mounts.

The tunnel contains hardware essential to final docking operations. This includes connectors for the electrical umbilicals, docking latches, probe-mounting lugs, tunnel lights, and deadfacing switches.

The probe assembly provides initial CSM-LM coupling and attenuates impact energy imposed by vehicle contact. The probe assembly may be folded for removal and for stowage within either end of the CSM transfer tunnel.

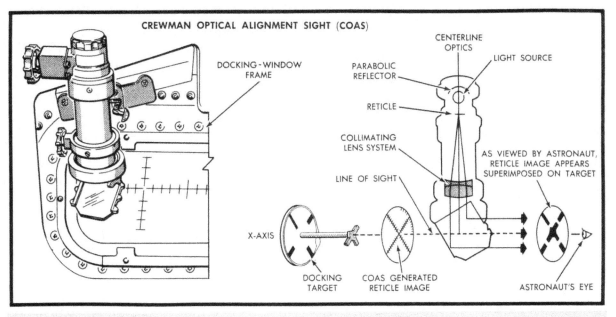

CREWMAN OPTICAL ALIGNMENT SIGHT (COAS)

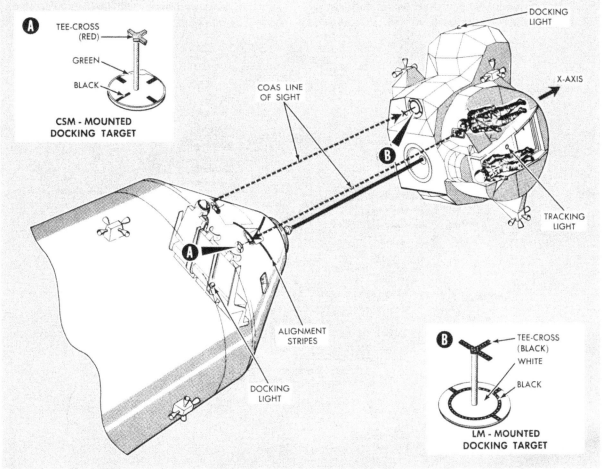

R-26

Docking Aids

CREW MISCELLANEOUS EQUIPMENT

Miscellaneous equipment required for completion of crew operations consists of in-flight data with checklists, emergency tool B, and window shades.

The in-flight data are provided in a container in the left-hand midsection. The Commander's checklist is stowed at his station. The in-flight data kit is stowed in a stowage compartment. The packages include the flight plan, experiments data and checklist, mission log and data book, systems data book and star charts.

Tool B (emergency wrench) is a modified Allen-head L-wrench. It is 6.25 inches long and has a 4.250-inch drive shaft with a 7/10-inch drive. The wrench can apply a torque of 4,175 inch-pounds; it has a ball-lock device to lock the head of the drive shaft. The wrench is stowed on the right side

stowage area inside the cabin. It is a contingency tool for use with the probe and drogue, and for opening the CM hatch from outside.

Window shades are used for the overhead (docking) window and forward windows. The window shade material is Aclar. The surface facing outside the cabin has a highly reflective metallic coating. The shade is secured at the bottom (rolled position). To cover the window, the shade is unrolled, flattened against the frame area and secured with snap fasteners.

MODULARIZED EQUIPMENT STOWAGE ASSEMBLY

The ME3A pallet is located in quad 4 of the descent stage. The pallet is deployed by the extra-vehicular astronaut when the LM is on the lunar surface. It contains fresh PLSS batteries and LiOH

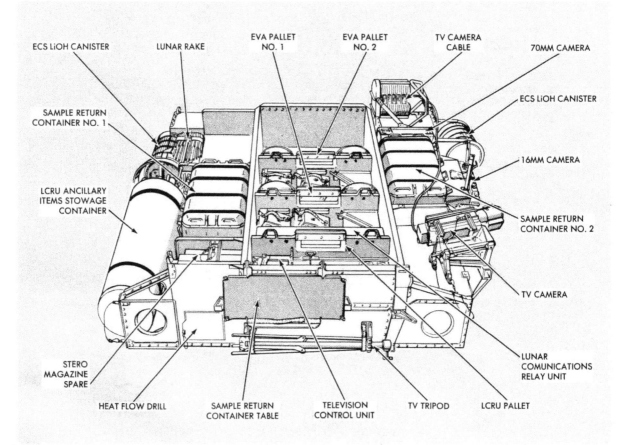

Modularized Equipment Stowage Assembly

cartridges, a TV camera and cable, still camera, tools for obtaining lunar geological samples, food, film, and containers in which to store the samples. It also has a folding table on which to place the sample return containers. Pallets are provided and are used to transfer the PLSS batteries and the cartridges to the cabin.

QUAD 3 PALLET ASSEMBLY

The quad 3 pallet assembly contains two pallets, a Lunar Roving Vehicle (LRV) pallet, and a pallet holding the Lunar Retro-Ranging Reflector. The LRV pallet contains a lunar geological exploration tool carrier, a lunar dust brush, a gnomen, a recording penetrometer, tongs, a trenching tool, collection bags, and other items needed during lunar exploration.

APOLLO LUNAR SURFACE EXPERIMENT PACKAGE

The Apollo Lunar Surface Experiment Package (ALSEP) consists of two packages of scientific instruments and supporting subsystems capable of transmitting scientific data to earth for one year. These data will be used to derive information regarding the composition and structure of the lunar body, its magnetic field, atmosphere and solar wind. Two packages are stowed in quad 2 of the descent stage. The packages are deployed on the lunar surface by the extravehicular astronaut.

ALSEP power is supplied by a radioisotope thermoelectric generator (RTG). Electrical energy is developed through thermoelectric action. The RTG provides a minimum of 16 volts at 56.2 watts to a power-conditioning unit. The radioisotopes fuel capsule emits nuclear radiation and approximately 1,500 thermal watts continuously. The surface temperature of the fuel capsule is approximately 1,400° F. The capsule is stowed in a graphite cask, which is externally mounted on the descent stage. The capsule is removed from the cask and installed in the RTG.

LASER RANGING RETRO-REFLECTOR

The laser ranging retro-reflector is a passive experiment with an array of optical reflectors that serve as targets for laser-pointing systems on earth. The experiment is designed to accurately measure the distance between earth and the moon.

ASTRONAUT PLACING EQUIPMENT TRANSFER BAG ON MESA TABLE

ASTRONAUT STOWING LiOH CARTRIDGE IN EQUIPMENT TRANSFER BAG

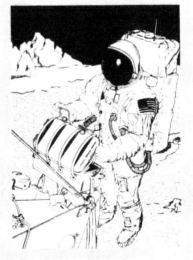

ASTRONAUT PLACING LUNAR SAMPLE IN SAMPLE RETURN CONTAINER

R-28

Application of MESA on Lunar Surface

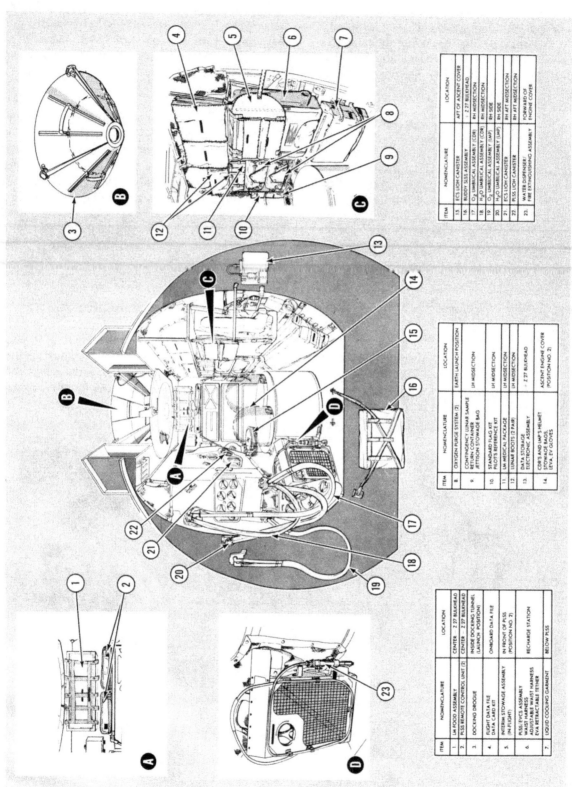

ITEM	NOMENCLATURE	LOCATION
15	ECS LIOH CANISTER	AFT OF ASCENT COVER
16	BUDDY SLSS ASSEMBLY	Z 27 BULKHEAD
17	O₂ UMBILICAL ASSEMBLY (CDR)	RH MIDSECTION
18	H₂O UMBILICAL ASSEMBLY (CDR)	RH MIDSECTION
19	O₂ UMBILICAL ASSEMBLY (LMP)	RH SIDE
20	H₂O UMBILICAL ASSEMBLY (LMP)	RH SIDE
21	ECS LIOH CANISTER	RH AFT MIDSECTION
22	PLSS LIOH CANISTER	RH AFT MIDSECTION
23	WATER DISPENSER/FIRE EXTINGUISHING ASSEMBLY	FORWARD OF ENGINE COVER

ITEM	NOMENCLATURE	LOCATION
8	OXYGEN PURGE SYSTEM (2)	EARTH LAUNCH POSITION
9	CONTINGENCY LUNAR SAMPLE RETURN CONTAINER JETTISON STOWAGE BAG	LH MIDSECTION
10	STANDARD FLAG KIT PILOT'S REFERENCE KIT	LH MIDSECTION
11	LM MEDICAL PACKAGE	LH MIDSECTION
12	LUNAR BOOTS (2 PAIR)	LH MIDSECTION
13	DATA STORAGE ELECTRONIC ASSEMBLY	Z 27 BULKHEAD
14	CDR'S AND LMP'S HELMET STOWAGE BAGS, LEVA, EV GLOVES	ASCENT ENGINE COVER (POSITION NO 2)

ITEM	NOMENCLATURE	LOCATION
1	LM FOOD ASSEMBLY	CENTER Z 27 BULKHEAD
2	PLSS REMOTE CONTROL UNIT (2)	CENTER Z 27 BULKHEAD
3	DOCKING DROGUE	INSIDE DOCKING TUNNEL (LAUNCH POSITION)
4	FLIGHT DATA FILE DATA CARD KIT	ONBOARD DATA FILE
5	INTERIM STOWAGE ASSEMBLY (IN FLIGHT)	IN FRONT OF PLSS (POSITION NO 2)
6	PLSS: EVCS ASSEMBLY WAIST HARNESS ADJUSTABLE WAIST HARNESS EVA RETRACTABLE TETHER	RECHARGE STATION
7	LIQUID COOLING GARMENT	BELOW PLSS

Crew Miscellaneous Equipment (Sheet 1)

R-29A

GRUMMAN

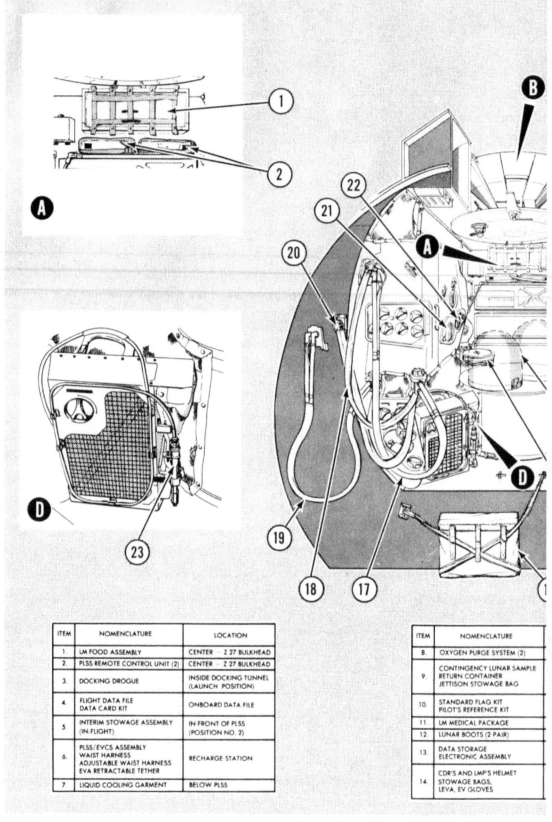

ITEM	NOMENCLATURE	LOCATION
1.	LM FOOD ASSEMBLY	CENTER – Z 27 BULKHEAD
2.	PLSS REMOTE CONTROL UNIT (2)	CENTER – Z 27 BULKHEAD
3.	DOCKING DROGUE	INSIDE DOCKING TUNNEL (LAUNCH POSITION)
4.	FLIGHT DATA FILE DATA CARD KIT	ONBOARD DATA FILE
5.	INTERIM STOWAGE ASSEMBLY (IN-FLIGHT)	IN FRONT OF PLSS (POSITION NO. 2)
6.	PLSS/EVCS ASSEMBLY WAIST HARNESS ADJUSTABLE WAIST HARNESS EVA RETRACTABLE TETHER	RECHARGE STATION
7.	LIQUID COOLING GARMENT	BELOW PLSS

ITEM	NOMENCLATURE
8.	OXYGEN PURGE SYSTEM (2)
9.	CONTINGENCY LUNAR SAMPLE RETURN CONTAINER JETTISON STOWAGE BAG
10.	STANDARD FLAG KIT PILOT'S REFERENCE KIT
11.	LM MEDICAL PACKAGE
12.	LUNAR BOOTS (2 PAIR)
13.	DATA STORAGE ELECTRONIC ASSEMBLY
14.	CDR'S AND LMP'S HELMET STOWAGE BAGS, LEVA, EV GLOVES

Crew Miscellaneous E

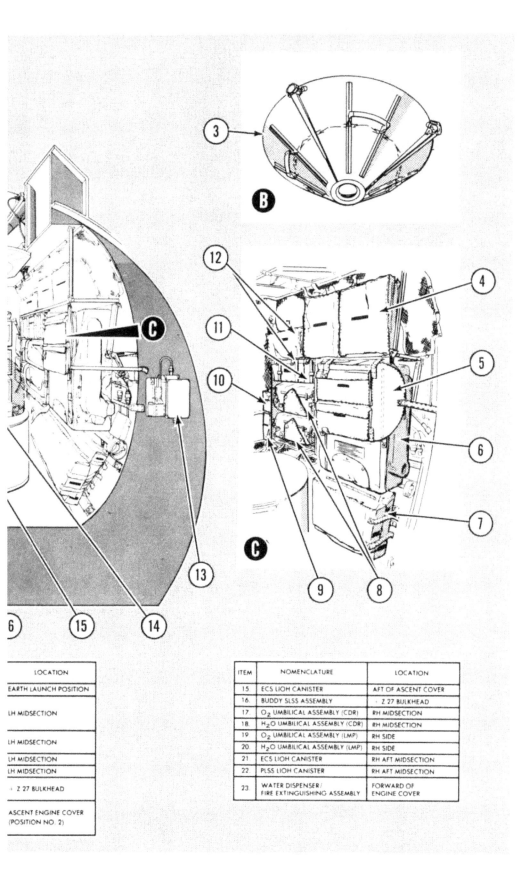

	LOCATION
	EARTH LAUNCH POSITION
	LH MIDSECTION
	LH MIDSECTION
	LH MIDSECTION
	LH MIDSECTION
	· Z 27 BULKHEAD
	ASCENT ENGINE COVER (POSITION NO. 2)

ITEM	NOMENCLATURE	LOCATION
15	ECS LIOH CANISTER	AFT OF ASCENT COVER
16.	BUDDY SLSS ASSEMBLY	· Z 27 BULKHEAD
17.	O$_2$ UMBILICAL ASSEMBLY (CDR)	RH MIDSECTION
18.	H$_2$O UMBILICAL ASSEMBLY (CDR)	RH MIDSECTION
19	O$_2$ UMBILICAL ASSEMBLY (LMP)	RH SIDE
20	H$_2$O UMBILICAL ASSEMBLY (LMP)	RH SIDE
21	ECS LIOH CANISTER	RH AFT MIDSECTION
22.	PLSS LIOH CANISTER	RH AFT MIDSECTION
23	WATER DISPENSER / FIRE EXTINGUISHING ASSEMBLY	FORWARD OF ENGINE COVER

quipment (Sheet 1)

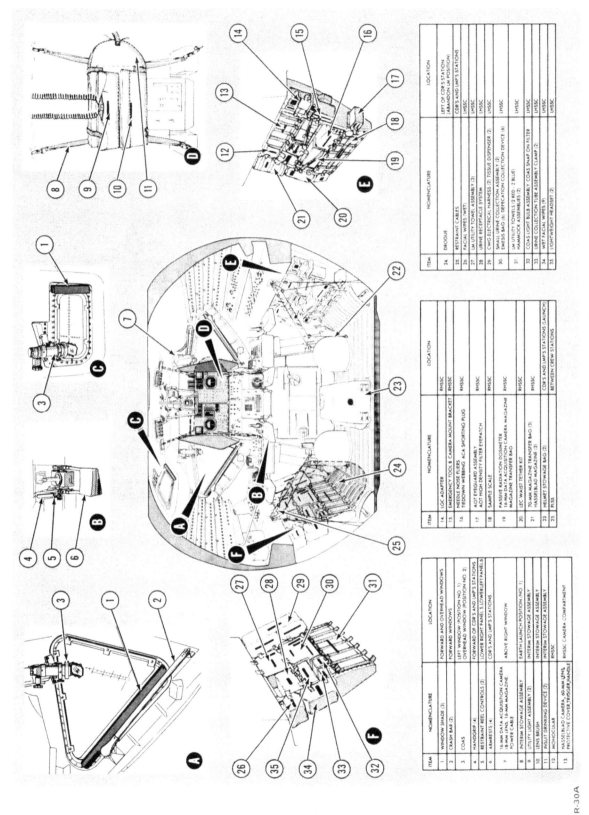

Crew Miscellaneous Equipment (Sheet 2)

ITEM	NOMENCLATURE	LOCATION
24	DROGUE	LEFT OF CDR'S STATION (ABANDON LM POSITION)
25	RESTRAINT CABLES	CDR'S AND LMP'S STATIONS
26	FACIAL WIPES (WET)	LHSSC
27	LM UTILITY TOWEL ASSEMBLY (3)	LHSSC
28	URINE RECEPTACLE SYSTEM	LHSSC
29	CWG ELECTRICAL HARNESS (2); TISSUE DISPENSER (2)	LHSSC
30	SMALL URINE COLLECTION ASSEMBLY (2); EMESIS BAG (6) DEFECATION COLLECTION DEVICE (6)	LHSSC
31	LM UTILITY TOWELS (2 RED - 2 BLUE); HAMMOCK ASSEMBLIES (2)	LHSSC
32	COAS LIGHT BULB ASSEMBLY COAS SNAP ON FILTER	LHSSC
33	URINE COLLECTION TUBE ASSEMBLY CLAMP (2)	LHSSC
34	WET FACIAL WIPES (9)	LHSSC
35	LIGHTWEIGHT HEADSET (2)	LHSSC

ITEM	NOMENCLATURE	LOCATION
14	LOC ADAPTER	RHSSC
15	EMERGENCY TOOL & CAMERA MOUNT BRACKET	RHSSC
16	NEEDLE NOSE PLIERS; TIEDOWN WEBBING ACA SHORTING PLUG	RHSSC
17	AOT EYEGUARD ASSEMBLY AOT HIGH DENSITY FILTER EYEPATCH	RHSSC
18	SAMPLE SCALE	RHSSC
19	PASSIVE RADIATION DOSIMETER 16-MM DATA ACQUISITION CAMERA MAGAZINE MAGAZINE TRANSFER BAG	RHSSC
20	LEC WAIST TETHER KIT	RHSSC
21	70-MM MAGAZINE TRANSFER BAG (3); HASSELBLAD MAGAZINE (3)	RHSSC
22	HELMET STOWAGE BAG (2)	CDR'S AND LMP'S STATIONS (LAUNCH)
23	PLSS	BETWEEN CREW STATIONS

ITEM	NOMENCLATURE	LOCATION
1	WINDOW SHADE (3)	FORWARD AND OVERHEAD WINDOWS
2	CRASH BAR (2)	FORWARD WINDOWS
3	COAS	LEFT WINDOW (POSITION NO 1) OVERHEAD WINDOW (POSITION NO 2)
4	HANDGRIP (4)	FORWARD OF CDR'S AND LMP'S STATIONS
5	RESTRAINT REEL CONTROLS (2)	LOWER RIGHT PANEL 3; LOWER LEFT PANEL 6
6	ARMRESTS (4)	CDR'S AND LMP'S STATIONS
7	16-MM DATA ACQUISITION CAMERA 16-MM LENS, 16-MM MAGAZINE, POWER CABLE	ABOVE RIGHT WINDOW
8	INTERIM STOWAGE ASSEMBLY	EARTH LAUNCH POSITION (NO 1)
9	UTILITY LIGHT ASSEMBLY (2)	INTERIM STOWAGE ASSEMBLY
10	LENS BRUSH	INTERIM STOWAGE ASSEMBLY
11	INSUIT DRINKING DEVICE (2)	RHSSC
12	MONOCULAR	RHSSC CAMERA COMPARTMENT
13	HASSELBLAD CAMERA, 60-MM LENS, PROTECTIVE COVER, TRIGGER HANDLE	

R-30A

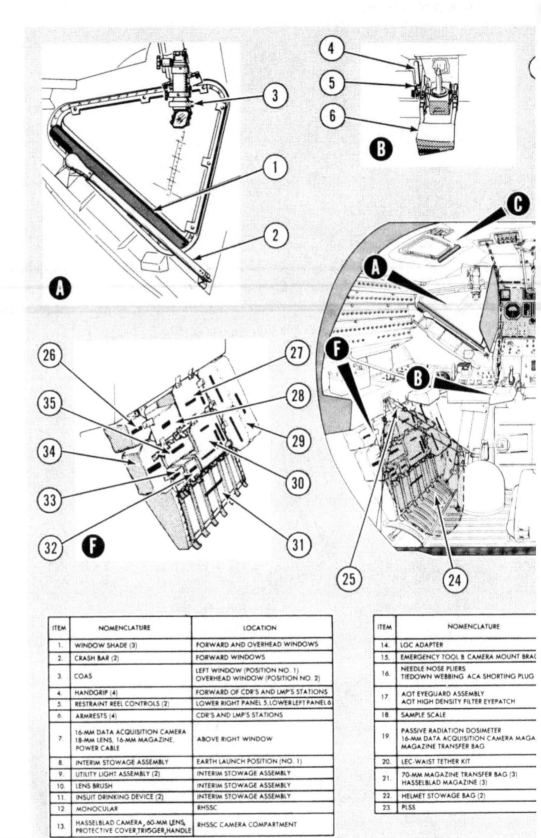

ITEM	NOMENCLATURE	LOCATION
1.	WINDOW SHADE (3)	FORWARD AND OVERHEAD WINDOWS
2.	CRASH BAR (2)	FORWARD WINDOWS
3.	COAS	LEFT WINDOW (POSITION NO. 1) OVERHEAD WINDOW (POSITION NO. 2)
4.	HANDGRIP (4)	FORWARD OF CDR'S AND LMP'S STATIONS
5.	RESTRAINT REEL CONTROLS (2)	LOWER RIGHT PANEL 5, LOWER LEFT PANEL 6
6.	ARMRESTS (4)	CDR'S AND LMP'S STATIONS
7.	16-MM DATA ACQUISITION CAMERA 18-MM LENS, 16-MM MAGAZINE, POWER CABLE	ABOVE RIGHT WINDOW
8.	INTERIM STOWAGE ASSEMBLY	EARTH LAUNCH POSITION (NO. 1)
9.	UTILITY LIGHT ASSEMBLY (2)	INTERIM STOWAGE ASSEMBLY
10.	LENS BRUSH	INTERIM STOWAGE ASSEMBLY
11.	INSUIT DRINKING DEVICE (2)	INTERIM STOWAGE ASSEMBLY
12.	MONOCULAR	RHSSC
13.	HASSELBLAD CAMERA, 60-MM LENS, PROTECTIVE COVER, TRIGGER, HANDLE	RHSSC CAMERA COMPARTMENT

ITEM	NOMENCLATURE
14.	LGC ADAPTER
15.	EMERGENCY TOOL B CAMERA MOUNT BRAC
16.	NEEDLE NOSE PLIERS TIEDOWN WEBBING ACA SHORTING PLUG
17.	AOT EYEGUARD ASSEMBLY AOT HIGH DENSITY FILTER EYEPATCH
18.	SAMPLE SCALE
19.	PASSIVE RADIATION DOSIMETER 16-MM DATA ACQUISITION CAMERA MAGA MAGAZINE TRANSFER BAG
20.	LEC-WAIST TETHER KIT
21.	70-MM MAGAZINE TRANSFER BAG (3) HASSELBLAD MAGAZINE (3)
22.	HELMET STOWAGE BAG (2)
23.	PLSS

Crew Miscellaneous Eq

GRUMMAN

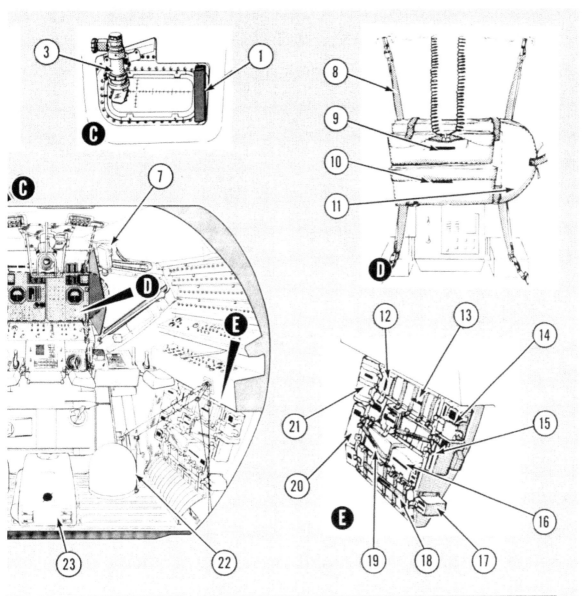

ITEM	NOMENCLATURE	LOCATION
		RHSSC
	OUNT BRACKET	RHSSC
	ING PLUG	RHSSC
	TCH	RHSSC
		RHSSC
	ERA MAGAZINE	RHSSC
		RHSSC
	G (3)	RHSSC
		CDR'S AND LMP'S STATIONS (LAUNCH)
		BETWEEN CREW STATIONS

ITEM	NOMENCLATURE	LOCATION
24.	DROGUE	LEFT OF CDR'S STATION (ABANDON LM POSITION)
25.	RESTRAINT CABLES	CDR'S AND LMP'S STATIONS
26.	FACIAL WIPES (WET)	LHSSC
27.	LM UTILITY TOWEL ASSEMBLY (3)	LHSSC
28.	URINE RECEPTACLE SYSTEM	LHSSC
29.	CWG ELECTRICAL HARNESS (2) TISSUE DISPENSER (2)	LHSSC
30.	SMALL URINE COLLECTION ASSEMBLY (2) EMESIS BAG (6) DEFECATION COLLECTION DEVICE (6)	LHSSC
31.	LM UTILITY TOWELS (2 RED - 2 BLUE) HAMMOCK ASSEMBLIES (2)	LHSSC
32.	COAS LIGHT BULB ASSEMBLY COAS SNAP ON FILTER	LHSSC
33.	URINE COLLECTION TUBE ASSEMBLY CLAMP (2)	LHSSC
34.	WET FACIAL WIPES (9)	LHSSC
35.	LIGHTWEIGHT HEADSET (2)	LHSSC

us Equipment (Sheet 2)

ENVIRONMENTAL CONTROL

QUICK REFERENCE DATA

Atmosphere revitalization section (ARS)

Cabin pressure	4.8±0.2 psia (normal, steady-state)
Suit circuit pressure	
Cabin mode	4.8±0.2 psia (may exceed 5.0 psia during powered flight)
Egress mode	3.8±0.2 psia (4.7 psia maximum during powered flight)
Suit inlet temperature range	
Gas cooled mode	With suit temperature control valve in full cold position, temperature ranges from +38° to +65° F; in full hot position, from +42° to 100° F.
Liquid cooled mode	Within 7° of HTS glycol temperature at liquid circuit cooling package inlet.
Relative humidity	40% to 80% (during periods of cabin mode operation)
Suit circuit fan flow	At 4.8 psia - 36.0 pounds per hour (minimum)
	At 3.8 psia - 28.4 pounds per hour (minimum)
Liquid cooled garment flow	At 5.65 psid - 4.0 pounds per minute (nominal) (1.9 minimum each suit)

Oxygen supply and cabin pressure control section (OSCPCS)

Suit pressure increase	First 1-psi increase may occur in less than 1 second. Each succeeding 1-psi increase occurs in not less than 8 seconds.
Cabin repressurization and emergency oxygen valve delivery rate	
Descent mode	4 pounds per minute (maximum)
Ascent mode	8 pounds per minute at 500 psia
PLSS refill	1.60 pound/refill at 1410 psia; can only be partially filled at lower pressures
Cabin volume	195 cubic feet (nominal)
Manual cabin pressure dump rate (without oxygen inflow)	Each dump valve from 5.0 psia down to 0.08 psia in 180 seconds; both valves, 90 seconds
Number of cabin repressurizations	4 at 5.5 pounds each (nominal)
Cabin repressurization time	1 minute (minimum) - 6 minutes (maximum)
Cabin pressure switch settings	When cabin pressure drops to 3.7 to 4.45 psia, contacts close. When cabin pressure increases to 4.40 to 5.0 psia, contacts open.
Descent oxygen tanks	
Capacity	48.01 pounds (each tank) at 2,690 psia and +70° F (residual oxygen: 84 pound (each tank) at 50 psia and +70° F)
Burst pressure	4700 psig

Oxygen supply and cabin pressure control section (cont)

Ascent oxygen tanks
Capacity

2.43 pounds (each tank) at 840 psia and +70° F (residual oxygen: 0.14 pounds (each tank) at 50 psia and +70° F)

Burst pressure

1,500 psig

Bypass relief valve
Full flow pressure
Cracking pressure
Reseat pressure

3,030 psig (maximum) at +75° F
2,875 psig (minimum) at +75° F
2,850 psig (minimum) at +75° F

Overboard relief valve
Full flow pressure
Cracking pressure
Reseat pressure

1,090 psig (maximum) at +75° F
1,025 psig (minimum) at +75° F
985 psig (minimum) at +75° F

High-pressure regulator
Outlet pressure with primary and secondary regulators operating normally

At inlet pressure of 1,100 to 3,000 psig and flow of 0.01 to 4.0 pounds per hour, and inlet pressure of 975 to 1,100 psig and flow of 0.1 pound per hour, will regulate to 875 to 960 psig at 75° F

PLSS Oxygen Regulator
Regulator outlet pressure with primary and secondary regulators operating normally

1395 psia (minimum) at +75° F
1445 psia (maximum) at +75° F

Overboard Relief Valve
Full flow pressure
Cracking pressure
Reseat pressure

1560 psia (maximum) at +75° F
1500 psia (minimum) at +75° F
1460 psia (minimum) at +75° F

Flow Limiter

1.5 lbs/hr (maximum) at 2700 psia inlet and 1500 psia outlet

Water management section (WMS)

Descent water tanks
Capacity
Initial fill pressure
Residual water
Pressure upon expulsion of all expellable water

333 pounds (each tank) at 0.75 fill ratio
48.2 psia (maximum) at +80° F
6.66 pounds
11.0 psia (minimum) at +35° F

Ascent water tanks
Capacity
Initial fill pressure
Residual water

42.5 pounds (each tank) at 0.75 fill ratio
48.2 psia (maximum) at +80° F
0.85 pound (each tank)

Pressure regulators discharge pressure

0.4 to 1.0 psi above ARS gas pressure

PLSS refill (each)

11.8 pounds of water (maximum)

Heat transport section (HTS)

Coolant

Solution of ethylene glycol and water (35% and 65%, respectively, by weight) with inhibitors

Approximately 25 pounds of coolant is used in HTS.

Coolant slush point

-3° F

Coolant pump rated flow

Flow rate of 222 pounds per hour (minimum) at +40° F, 30 psid, and 28-vdc input voltage

Heat transport section (HTS) (cont)

Coolant pump bypass relief valve
    Cracking pressure   33 psi (minimum)
    Fully open pressure   39 psi (maximum)
    Reseat pressure   1 psi less than cracking

Coolant temperatures   +29° to +120° F
    Flow into primary sublimator   +38° to 100° F
    Flow out of sublimator   +29° to +36° F
    Flow into secondary sublimator   +60.1° F (nominal)
    Flow out of sublimator   +47° F (nominal)

Vacuum fill requirements   HTS withstands internal pressure of 500 microns in sea-level ambient pressure environment.

Primary and secondary coolant filters
    Efficiency   35 microns absolute for primary; 45 microns absolute for secondary
    Maximum bypass valve cracking pressure   0.4 psid for primary; 1.0 psid for secondary

Coolant flow
    Primary coolant loop   222 pounds per hour (minimum)
    Secondary coolant loop   222 pounds per hour (minimum)

Cold plates
    Coolant inlet operating temperature   +32° to +100° F
    Coolant inlet operating pressure   5 to 45 psia
    Coolant inlet operating flow   12 to 85 pounds per hour, depending on cold-plate size

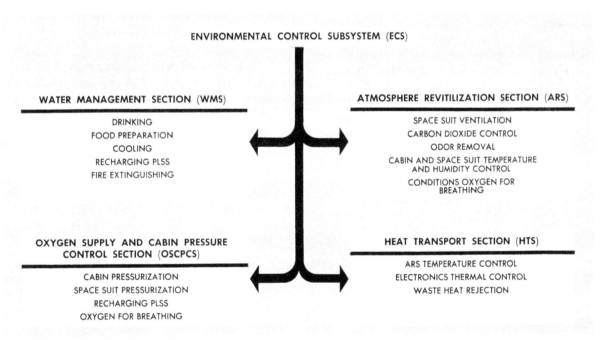

ENVIRONMENTAL CONTROL SUBSYSTEM (ECS)

WATER MANAGEMENT SECTION (WMS)
DRINKING
FOOD PREPARATION
COOLING
RECHARGING PLSS
FIRE EXTINGUISHING

ATMOSPHERE REVITILIZATION SECTION (ARS)
SPACE SUIT VENTILATION
CARBON DIOXIDE CONTROL
ODOR REMOVAL
CABIN AND SPACE SUIT TEMPERATURE AND HUMIDITY CONTROL
CONDITIONS OXYGEN FOR BREATHING

OXYGEN SUPPLY AND CABIN PRESSURE CONTROL SECTION (OSCPCS)
CABIN PRESSURIZATION
SPACE SUIT PRESSURIZATION
RECHARGING PLSS
OXYGEN FOR BREATHING

HEAT TRANSPORT SECTION (HTS)
ARS TEMPERATURE CONTROL
ELECTRONICS THERMAL CONTROL
WASTE HEAT REJECTION

R-31

Block Diagram of the Environmental Control Subsystem

The Environmental Control Subsystem (ECS) enables pressurization of the cabin and space suits, controls the temperature of electronic equipment, and provides breathable oxygen for the astronauts. It also provides water for drinking, cooling, fire extinguishing, and food preparation and supplies oxygen and water to the portable life support system (PLSS).

The major portion of the ECS is in the cabin. The peripheral ECS equipment, such as oxygen and water tanks, is located outside the cabin, in the ascent and descent stages. The ECS consists of the following sections:

Atmosphere revitalization section (ARS)

Oxygen supply and cabin pressure control section (OSCPCS)

Water management section (WMS)

Heat transport section (HTS)

The ARS purifies and conditions the oxygen for the cabin and the space suits. Oxygen conditioning consists of removing carbon dioxide, odors, particulate matter, and excess water vapor. It also temperature conditions the oxygen and water used to cool the astronauts' space suits.

The OSCPCS stores gaseous oxygen and maintains cabin and suit pressure by supplying oxygen to the ARS to compensate for crew metabolic consumption and cabin or suit leakage. Two oxygen tanks in the descent stage provide oxygen during descent and lunar stay. Two oxygen tanks in the ascent stage are used during ascent and rendezvous.

The WMS supplies water for drinking, cooling, fire extinguishing, and food preparation, and for refilling the PLSS cooling water tank. It also provides for delivery of water from ARS water separators to HTS sublimators and from water tanks to ARS and HTS sublimators.

The water tanks are pressurized before launch, to maintain the required pumping pressure in the tanks. The descent stage tanks supply most of the water required until staging occurs. After staging, water is supplied by the two ascent stage tanks. A self-sealing valve delivers water for drinking and food preparation.

The HTS consists of a primary coolant loop and a secondary coolant loop. The secondary loop serves as a backup loop; it functions if the primary loop fails. A water-glycol solution circulates through each loop. The primary loop provides temperature control for batteries, electronic equipment that requires active thermal control, and for the oxygen and water that circulates through the space suits. The batteries and electronic equipment are mounted on cold plates and rails through which coolant is routed to remove waste heat. The cold plates used for equipment that is required for mission abort contain two separate coolant passages: one for the primary loop and one for the secondary loop. The secondary coolant loop, which is used only if the primary loop is inoperative, serves only these cold plates.

In-flight waste heat rejection from both coolant loops is achieved by the primary and secondary sublimators, which are vented overboard. A coolant pump recirculation assembly contains all the HTS coolant pumps and associated check and relief valves. Coolant flow from the assembly is directed through parallel circuits to the cold plates for the electronic equipment and the oxygen-to-glycol and water-to-glycol heat exchangers in the ARS.

FUNCTIONAL DESCRIPTION

The functional description of each of the four major ECS sections is supported by a functional flow diagram, which, to reduce complexity, does not contain electrical circuitry.

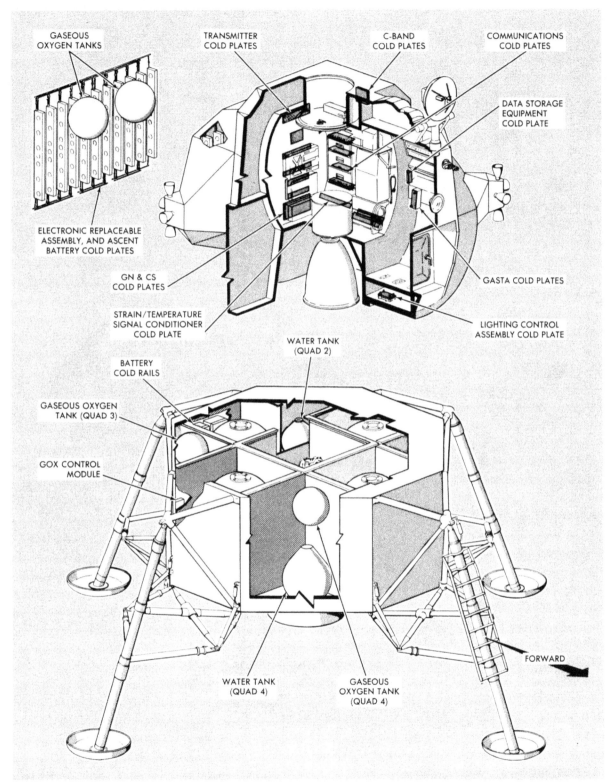

Environmental Control Subsystem, Component Location (Sheet 1)

R-32A

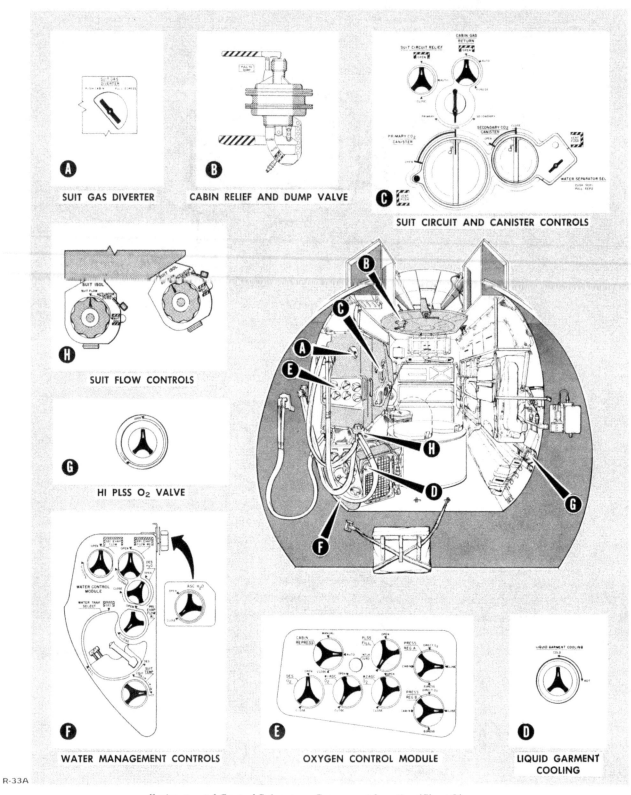

A SUIT GAS DIVERTER

B CABIN RELIEF AND DUMP VALVE

C SUIT CIRCUIT AND CANISTER CONTROLS

H SUIT FLOW CONTROLS

G HI PLSS O₂ VALVE

F WATER MANAGEMENT CONTROLS

E OXYGEN CONTROL MODULE

D LIQUID GARMENT COOLING

R-33A

Environmental Control Subsystem, Component Location (Sheet 2)

GRUMMAN

ATMOSPHERE REVITALIZATION SECTION

The ARS is a recirculation system that conditions oxygen by cooling or heating, dehumidifying, and deodorizing it for use within the space suits and cabin, circulates water through the liquid cooling garment to provide cooling during peak heat loads, and recirculates and filters the cabin atmosphere. The major portion of the ARS is contained within the suit circuit assembly.

In normal operation, conditioned oxygen flows to the space suits and is discharged through the return umbilical to the suit circuit. Suit circuit pressure, sensed at a point downstream of the suits, is referenced to the oxygen regulators that control pressure by supplying makeup oxygen to the suit circuit. The suit circuit relief valve protects the suit circuit against overpressurization, by venting the ARS.

The cabin position of the suit gas diverter valve is used during pressurized-cabin operation, to direct sufficient conditioned oxygen to the cabin to control carbon dioxide and humidity levels. Pulling the valve handle selects the egress position to isolate the suit circuit from the cabin. The egress position is used for all unpressurized-cabin operations, as well as closed suit mode with pressurized cabin. An electrical solenoid override automatically repositions the valve from cabin to egress when cabin pressure drops below the normal level or when the egress position is selected on the pressure regulators.

With the suit gas diverter valve set to the cabin position, cabin discharge oxygen is returned to the suit circuit through the cabin gas return valve. Setting the cabin gas return valve to automatic position enables cabin pressure to open the valve. When the cabin is depressurized, differential pressure closes the valve, preventing suit pressure loss.

A small amount of oxygen is tapped from the suit circuit upstream of the PGA inlets and fed to the carbon dioxide partial pressure sensor, which provides a voltage to the CO_2 partial pressure indicator.

The primary and secondary carbon dioxide and odor removal canisters are connected to form a parallel loop. The primary canister contains a LM cartridge with a capacity of 41 man hours; the secondary canister, a PLSS cartridge with a capacity of 18 man hours. A debris trap in the primary canister cover prevents particulate matter from entering the cartridge. A relief valve in the primary canister permits flow to bypass the debris trap if it becomes clogged. Oxygen is routed to the CO_2 and odor removal canisters through the canister selector valve. The carbon dioxide level is controlled by passing the flow across a bed of lithium hydroxide (LiOH); odors are removed by absorption on activated charcoal. When carbon dioxide partial pressure reaches or exceeds 7.6 mm Hg, as indicated on the partial pressure CO_2 indicator, the CO_2 component caution light and ECS caution light go on. (The CO_2 component caution light also goes on when the CO_2 canister selector valve is in the secondary position.) The CO_2 canister selector valve is then set to the secondary position, placing the secondary canister onstream. The primary cartridge is replaced and the CO_2 canister selector valve is set to the primary position, placing the primary canister back onstream.

From the canisters, conditioned oxygen flows to the suit fan assembly, which maintains circulation in the suit circuit. Only one fan operates at a time. The ECS suit fan 1 circuit breaker is closed and the SUIT FAN selector switch is set to 1 to initiate suit fan operation. At startup, a fan differential pressure sensor is in the low position (low Δ P), which, through the fan condition signal control, energizes the ECS caution light and suit fan component caution light. The lights remain on until the differential pressure across the operating fan increases sufficiently to cause the differential pressure sensor to move to the normal position. If the differential pressure drops to 6.0 inches of water or less, the lights go on and switchover to fan No. 2 is required. The ECS caution light goes off when fan No. 2 is selected and the suit fan warning light goes on. The suit fan component caution light goes off when fan No. 2 comes up to speed and builds up normal differential pressure. The suit fan warning light and suit fan component caution light go off if fan No. 2 differential pressure reaches 9.0 inches of water. The fan check valve permits air to pass from the operating fan without backflow through the inoperative fan.

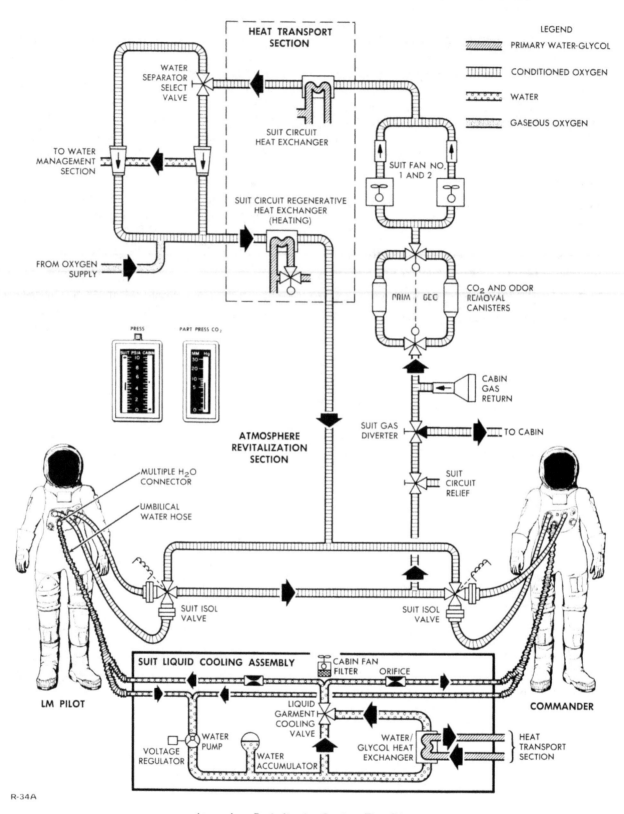

Atmosphere Revitalization Section, Flow Diagram

R-34A

From the check valve, the conditioned oxygen passes through a sublimator to the cooling heat exchanger. The sublimator cools the oxygen under emergency conditions. Heat transfer to the coolant in the heat exchanger reduces gas temperature and causes some condensation of water vapor.

The conditioned oxygen is next routed to two parallel-redundant water separators through the water separator selector valve. One separator, selected by pushing or pulling the water separator selector valve handle, is operated at a time. The separator is driven by the gas flowing through it. Moisture removed from the oxygen is discharged under a dynamic head of pressure sufficient to ensure positive flow from the separator to the WMS. A water drain carries residual water from the separators to a surge (collection) tank outside the recirculation system when the loop is shut down.

The conditioned oxygen from the water separator is mixed with makeup oxygen from the OSCPCS to maintain system pressure. The mixture flows through the regenerative (heating) heat exchanger, where the temperature may be increased, to the suit isolation valves. The suit temperature control valve on the water control module controls the flow of coolant through the regenerative heat exchanger. Setting the valve to the increase hot position increases oxygen temperature; setting it to decrease cold position reduces the temperature.

SUIT LIQUID COOLING ASSEMBLY

The suit liquid cooling assembly assists in removing metabolic heat by circulating cool water through the liquid cooled garment (LCG). Suit inlet temperature can be manually selected to within 7° F of the HTS glycol temperature at the assembly inlet. A pump maintains the flow of warm water returning from the LCG through the water umbilicals. An accumulator in the system compensates for volumetric changes and leakage. A mixing bypass valve controls the quantity of water that flows through the water-glycol heat exchanger. This bypassed (warm) water is mixed with the cool water downstream of the heat exchanger, flows through the water umbilicals back

to the LCG. The assembly also includes one cabin atmosphere recirculation fan.

OXYGEN SUPPLY AND CABIN PRESSURE CONTROL SECTION

The ECS descent stage oxygen supply hardware consists of the following: descent oxygen tanks, high-pressure fill coupling, high-pressure oxygen control assembly, interstage flex lines, PLSS recharge regulator assembly, and descent stage disconnects. The descent tank pressure transducers, part of the instrumentation subsystem, generate an output proportionate to tank pressure.

The ascent stage oxygen supply hardware consists of the following: ascent stage disconnects, interstage flex lines, oxygen module, two ascent oxygen tanks, a PLSS hose, PLSS oxygen disconnect, and the cabin pressure switch. Two automatic cabin pressure relief and dump valves, one in each hatch, are provided to protect the cabin from overpressurization. Two ascent stage tank pressure transducers and a selected oxygen supply transducer, part of the IS, operate in conjunction with the OSCPCS.

The OSCPCS stores gaseous oxygen, replenishes the ARS oxygen, and provides refills for the PLSS oxygen tank. Before staging, oxygen is supplied from the descent stage oxygen tanks. After staging, or if the descent tank is depleted, the ascent stage oxygen tanks supply oxygen to the oxygen control module. The high-pressure assembly in the descent stage, and the oxygen control module in the ascent stage, contain the valves and regulators necessary to control oxygen in the OSCPCS. The cabin relief and dump valves vent excess cabin pressure.

A high-pressure regulator reduces descent tank pressure, approximately 2,690 psia, to a level that is compatible with the components of the oxygen control module, approximately 900 psig. A series-redundant overboard relief valve protects the oxygen control module against excessive pressure caused by a defective regulator or by flow through the bypass relief valve. If the pressure on the outlet

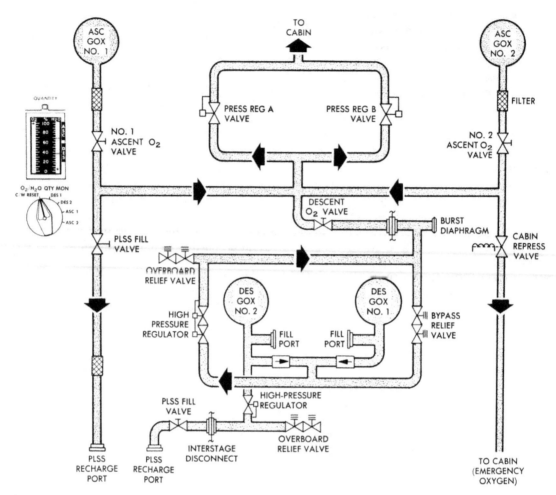

R-35A

Oxygen Supply and Cabin Pressure Control Section, Flow Diagram

side of the regulator rises to a dangerous level, the burst diaphragm assembly vents the high-pressure assembly to ambient. A poppet in the burst diaphragm assembly reseats when pressure in the high-pressure assembly is reduced to approximately 1,000 psig. Descent oxygen flow through the interstage disconnect to the oxygen control module is controlled with the descent oxygen shutoff valve. The interstage disconnect acts as a redundant seal to prevent loss of oxygen overboard after staging.

When ascent stage oxygen is required, the ascent oxygen shutoff valves are used to select their respective tank. A mechanical interlock prevents the valves from being opened unless the descent oxygen shutoff valve is closed. The mechanical interlock may be overridden (if the descent oxygen

shutoff valve cannot be closed and the ascent oxygen shutoff valves must be opened) by pressing the interlock override pushbutton on the oxygen control module.

From the oxygen shutoff valves, oxygen is routed to oxygen demand regulators, an alternate PLSS fill valve, and a cabin repressurization and emergency oxygen valve. For PLSS fills, oxygen is diverted from Descent Tank No. 2 upstream of the check valve and routed to the PLSS Oxygen Recharge Regulator Assembly. This series-redundant regulator reduces descent tank pressure to approximately 1400 psia. A series-redundant overboard relief valve protects the PLSS fill line against excessive pressure caused by a defective regulator. Oxygen flows through the interstage disconnect and a shutoff valve to the High Press PLSS fill disconnect. The PLSS fill disconnect connects the PLSS oxygen tank through a flexible service hose. A check valve in the PLSS disconnect is opened when the hose is connected. The valve automatically closes when the hose is disconnected. The oxygen demand regulators maintain the pressure of the suit circuit at a level consistent with normal requirements. Both regulators are manually controlled with a four-position handle; both are ordinarily set to the same position. The CABIN position is selected during normal pressurized-cabin operations, to provide oxygen at 4.8±0.2 psia. Setting the regulators to the egress position maintains suit circuit pressure at 3.8±0.2 psia. The direct O_2 position provides an unregulated flow of oxygen into the suit circuit. The close position shuts off all flow through the regulator. In the cabin and egress positions, the regulator is internally modulated by a reference pressure from the suit circuit. The demand regulators are redundant; either one can fulfill the ARS oxygen requirements.

If cabin pressure drops to 3.7 to 4.45 psia, the cabin pressure switch energizes the cabin repressurization valve and oxygen flows through the valve into the cabin. If cabin pressure builds up to 4.45 to 5.0 psia, the cabin pressure switch deenergizes the valve solenoid, shutting off the oxygen flow. The valve can maintain cabin pressure at 3.5 psia for at least 2 minutes following a 0.5-inch-diameter puncture of the cabin. It responds to signals from the cabin pressure switch during pressurized-cabin operation and to a suit circuit pressure switch during unpressurized operation. Manual override capabilities are provided.

Both cabin relief and dump valves (one in the forward hatch, the other in the overhead hatch) are manually and pneumatically operated. They prevent excessive cabin pressure and permit deliberate cabin depressurization. The valves automatically relieve cabin pressure when the cabin-to-ambient differential reaches 5.4 to 5.8 psid. When set to the automatic position, the valves can be manually opened with their external handle. The valves can each dump cabin pressure from 5.0 to 0.08 psia in 180 seconds without cabin inflow. In addition to relieving positive pressure, the valves relieve a negative cabin pressure condition.

To egress from the LM, the oxygen demand regulators are set to the egress position, turning off the cabin fan and closing the suit gas diverter valve; the cabin gas return valve is set to the egress position; and cabin pressure is dumped by opening the cabin relief and dump valve. When repressurizing the cabin, the cabin relief and dump valve is set to the automatic position, the oxygen demand regulator valves are set to the cabin position, and the cabin gas return valve is set to the automatic position. The cabin warning light goes on when the regulators are set to the cabin position and goes off when cabin pressure reaches the actuation pressure of the cabin pressure switch.

WATER MANAGEMENT SECTION

The WMS stores water for metabolic consumption, evaporative cooling, fire extinguishing, and PLSS water tank refill. It controls the distribution of this stored water and the water reclaimed from the ARS by the water separators. Reclaimed water is used only for evaporative cooling, in the ECS sublimators. Water is stored in two tanks in the descent stage and two smaller tanks in the upper midsection of the ascent stage. All four tanks are bladder-type vessels, which are pressurized with nitrogen before launch. The controls for the WMS are grouped together on the water control module located to the right rear of the LM Pilot's station.

Water from the descent stage water tanks is fed through a manually operated shutoff valve and a check valve to the PLSS water disconnect. Water quantity is determined by N_2 pressure tranducers. The output is displayed on the H_2O quantity indicator after the O_2/H_2O quantity monitor selector switch is set to the descent position. When the descent H_2O valve is opened, high-pressure water is available for drinking, food preparation, PLSS fill, and fire extinguishing.

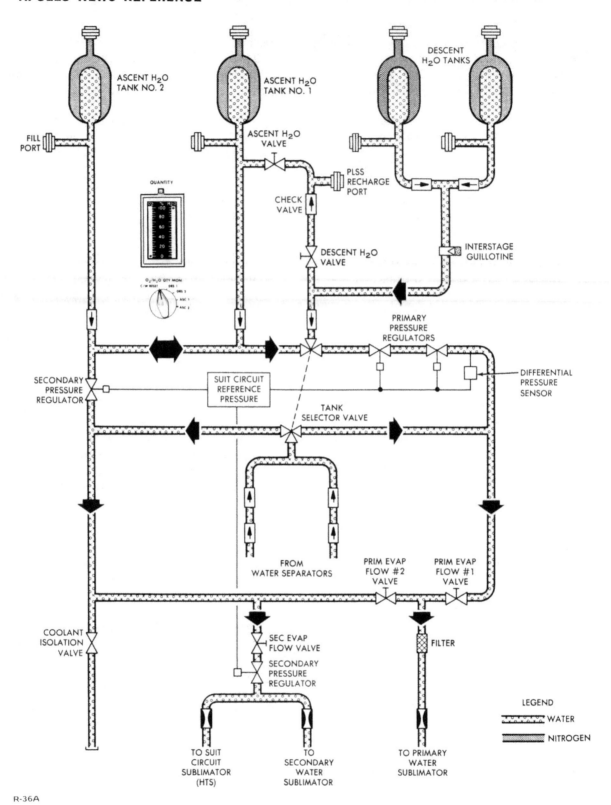

Water Management Section, Flow Diagram

R-36A

When the vehicle is staged, the descent interstage water feed line is severed by the interstage umbilical guillotine, and water is supplied from the ascent stage water tanks. Water quantity is monitored on either ascent water tank as required by switching. Water from ascent stage water tank No. 1 is fed to the PLSS water disconnect through the ascent water valve for drinking, food preparation, PLSS fill, and fire extinguishing.

Water from the four water tanks enters the water tank selector valve, which consists of two water diverting spools. Setting the valve to the descent or ascent position determines which tanks are on-line.

When using the descent tanks, water is supplied to the primary manifold (which consists of the primary pressure regulators and the primary evaporator flow No. 1 valve) by setting the water tank select valve to descent. The water flows through the series primary pressure regulators, which control water discharge pressure, in response to suit circuit gas reference pressure, at 0.4 to 1.0 psi above this gas pressure. With the primary evaporator flow valve opened, the water is routed to the primary sublimator. Discharge water from the water separator is routed through the secondary spool of the selector valve and joins the water from the primary pressure regulators. Setting the selector valve to ASC routes water from the ascent tanks through the primary pressure regulators and, with the primary evaporator flow No. 1 valve opened, to the primary sublimator. Water flow from the water separators is not changed by selection of the ASC position. If the primary pressure regulators fail, an alternative path to the primary sublimator is provided with the primary evaporator flow No. 2 valve opened. Water then flows directly from the ascent water tanks through the secondary pressure regulator and the primary evaporator flow No. 2 valve to the primary sublimator.

Under emergency conditions (failure of the primary HTS loop), water from the ascent tanks is directed through the secondary manifold (secondary pressure regulator) to the secondary sublimator and the suit circuit sublimator by opening the secondary evaporator flow valve. Discharge water from the water separators is also directed to the sublimator.

HEAT TRANSPORT SECTION

The HTS consists of two closed loops (primary and secondary) through which a water-glycol solution is circulated to cool the suit circuits and electronic equipment. Coolant is continuously circulated through cold plates and cold rails to remove heat from electronic equipment and batteries. Heat is removed by conduction and is rejected to space by process of sublimation. For the purpose of clarity, the primary and secondary coolant loops, and the primary and secondary coolant loop cold plates and rails are discussed separately in the following paragraphs.

The primary coolant loop is charged with coolant at the fill points and is then sealed. The glycol pumps force the coolant through the loop. The glycol accumulator maintains a constant head of pressure (5.25 to 9 psia, depending on coolant level) at the inlets of the primary loop glycol pumps. Coolant temperature at the inlets is approximately $+40^\circ$ F. A switch in a low-level sensor trips when only $10\pm5\%$ of coolant volume remains in the accumulator. When tripped, the switch provides a telemetry signal and causes the glycol caution light to go on.

The coolant is routed to the pumps through a filter. They are started by closing the appropriate circuit breakers and setting the glycol selector switch to pump 1 or pump 2. If the operating pump does not maintain a minimum differential pressure (ΔP) of 7 ± 2 psi, the ΔP switch generates a signal to energize the ECS caution light and the glycol component caution light. Selecting the other pump deenergizes the lights when the onstream pump develops a minimum ΔP of 5.0 to 9.0 psi. If both pumps fail, the secondary loop is activated by setting the water tank selector valve to the secondary, setting the glycol pump switch to INST

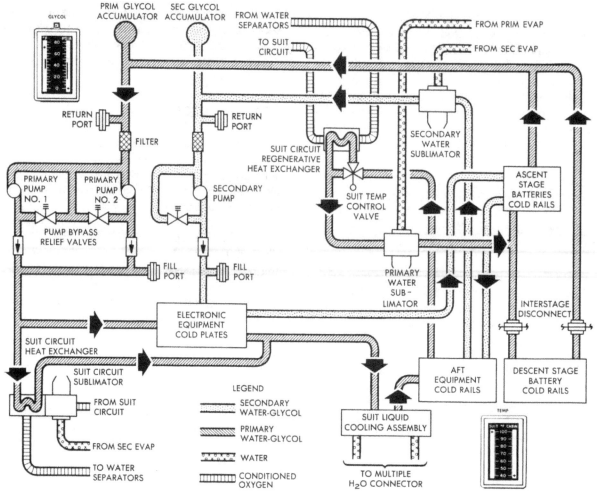

R-37

Heat Transport Section, Flow Diagram

(SEC), and closing the glycol pump secondary circuit breaker. Automatic transfer from primary pump No. 1 to primary pump No. 2 is initiated by closing the glycol automatic transfer circuit breaker and setting the selector switch to pump 1. When transfer is necessary, the caution lights go on, the transfer is accomplished, and the ECS caution light goes off. The glycol pump component light remains on.

If primary loop ΔP exceeds 33 psi, the pump bypass relief valve opens and routes the coolant back to the pump inlet, relieving the pressure. The valves start to open at 33 psi, are fully open at a maximum of 39 psia, and reseat at a minimum of 32 psia. Check valves prevent coolant from feeding back through an inoperative primary pump.

Part of the coolant leaving the recirculation assembly flows to the suit circuit heat exchanger to cool the suit circuit gas of the ARS. The remainder of the coolant flows to the electronic equipment mounted on cold plates. The flow paths then converge and the coolant is directed to the liquid cooling garment water glycol heat exchanger to cool suit water as required. The coolant then flows through the aft equipment bay cold rails.

A portion of the warmer coolant flow can be diverted to the suit circuit regenerative heat exchanger through the suit temperature control valve to increase suit inlet gas temperature. The diverted flow returning from the heat exchanger, combined with the bypassed coolant is routed to the primary sublimator.

The sublimator decreases the temperature of the coolant by rejecting heat to space through sublimation of water, followed by venting of generated steam through an overboard duct. Deflector plates, attached to the duct, prevent escaping steam from applying thrust to the vehicle. Water is fed to the sublimator at a pressure that exceeds 4.0 psia, but is less than 6.5 psia. The water pressure must be less than the suit circuit static pressure plus the head pressure from the water separators to the sublimator. The water regulators, referenced to suit circuit pressure, are in the water feed line to the sublimator. Regulated water pressure varies from 0.5 to 1.0 psid above suit circuit pressure. The sublimator inlet and outlet temperatures are sensed by temperature transducers, which provide telemetry signals. Coolant from the sublimator flows through the ascent and descent battery cold rails, then returns to the recirculation assembly.

Two self-sealing disconnects upstream and downstream of the glycol pumps permit servicing of the HTS. Interstage disconnects are installed in coolant lines that connect to the descent stage. Before staging, coolant flows through the ascent and descent stage battery cold rails. After staging, the interstage disconnects separate, the lines are sealed by spring-loaded valves, and the full coolant flow enters the ascent stage battery cold rails.

The secondary (emergency) coolant loop provides thermal control for those electronic assemblies and batteries whose performance is necessary to effect a safe return to the CSM. Cooling is provided by the secondary sublimator.

As in the primary loop, a secondary glycol accumulator provides pressure for the pump inlet side and compensates for loss due to leakage. A pump bypass relief valve relieves excessive pressure by routing coolant back to the pump inlet. A check valve at the discharge side of the glycol pump prevents coolant flow from bypassing the HTS during GSE operation. The coolant from the pump passes through the check valve to the secondary passage of the cold plates and cold rails of the electronics and batteries cold plate section. Waste heat is absorbed by the coolant. The warm coolant then flows to the secondary sublimator.

The secondary sublimator operates in the same manner as the primary sublimator in the primary coolant loop. Water for the sublimator is provided when the secondary evaporator flow valve is opened. The coolant returns to the pump for recirculation.

Equipment essential for mission abort is mounted on cold plates and rails that have two independent coolant passages, one for the primary loop and one for the secondary loop.

PRIMARY COOLANT LOOP COLD PLATES AND RAILS

The cold plates and rails in the primary coolant loop are arranged in three groups: upstream electronics, aft equipment bay, and batteries.

Coolant from the recirculation assembly flows into parallel paths that serve the upstream electronics cold plate group. In this group, the data storage electronics assembly (DSEA) is cooled by cold rails; the remainder of the electronics, by cold plates. The cold plates are in the pressurized and unpressurized areas of the LM. The flow rates through the parallel paths are controlled by flow restrictors, installed downstream of the cold plate group. The first upstream electronics flow path cools the suit circuit heat exchanger. The second flow path cools five cold plates mounted on the pressurized side of the equipment tunnel back wall. The third path serves the integrally cooled IMU and the rate gyro assembly (RGA) cold plate, both located in the unpressurized area (on the navigation base). The fourth path cools the abort sensor assembly (ASA) and pulse torque assembly (PTA) cold plates. All the plates for the fourth path are in the unpressurized area above the cabin; the ASA is on the navigation base of the alignment optical telescope (AOT). The fifth path serves the tracking light electronics (TLE), gimbal angle sequencing transformation assembly (GASTA), lighting control assembly (LCA), and DSEA plates: one in the unpressurized area in front of the cabin, a second one in the control and display panel area, a third one below the cabin floor, and another one on the left wall of the cabin.

The aft equipment bay is cooled by eight cold rails; the flow is in parallel. The batteries are cooled by cold rails. The ascent batteries are in the center section of the aft equipment bay; the descent batteries are on the -Z bulkhead of the descent stage. During the descent phase, the coolant flow is split between the descent batteries and the ascent batteries; the ascent batteries are not used during this time. When the stages are separated, quick-disconnects break the coolant lines and seal the ends; all coolant then flows through the ascent battery cold rails.

SECONDARY COOLANT LOOP COLD PLATES AND RAILS

The secondary coolant loop is for emergency use. Only cold plates and cold rails that have two independent passages (one for the primary loop and one for the secondary loop) are served by this loop.

In the upstream electronics area, the secondary coolant flow is split between three cold plates (RGA, ASA, and TLE) in parallel. The flow rate is controlled by a flow restrictor downstream of the TLE and RGA. After these three plates, the secondary loop cools the ascent battery cold rails and the aft equipment bay cold rails in a series-parallel arrangement. The coolant first flows through three ascent battery cold rails in parallel, then through eight aft equipment bay cold rails in parallel.

EQUIPMENT

ATMOSPHERE REVITALIZATION SECTION

SUIT CIRCUIT ASSEMBLY

Suit Circuit Relief Valve. The suit circuit relief valve protects the suit circuit against overpressurization. The valve has automatic, open, and close positions. Two externally mounted microswitches provide telemetry signals when the open or close position is selected.

In the automatic position, the valve responds to pressure sensed by the aneroid; it cracks open at approximately 4.3 psia to prevent overpressurization of the suit loop by allowing oxygen to flow to the cabin. At 4.7 psia, the valve is fully open and flows approximately 7.8 pph at +90° F. The valve reseats at approximately 4.3 psia. In the open position, the valve handle displaces the poppet from the seat to open the valve, regardless of pressure. In the close position, if the valve fails to reseat, the automatic poppet is left open, but an auxiliary poppet is closed, maintaining pressure.

Suit Gas Diverter Valve. The suit gas diverter valve is a manually operated, two-way valve (one inlet and two outlets) with a solenoid override in one direction. The valve is on the ECS package above the oxygen control module. When the valve handle is pushed into the cabin position, oxygen is directed into the cabin; pulling the valve handle to the egress position shuts off flow to the cabin.

An automatic closed-to-cabin feature is provided. If cabin pressure falls to below the normal level while the valve is set to CABIN, a solenoid is energized by the cabin pressure switch and the main spring returns the valve to the EGRESS position. Electrical power is also supplied to the diverter valve when the oxygen demand regulator valves are set to egress.

Cabin Gas Return Check Valve. The cabin gas return check valve is a spring-loaded, flapper-type valve. The valve has automatic, open, and egress (closed) positions. In the automatic position, the valve automatically permits cabin gas to return to the suit circuit. When the cabin is depressurized, the suit circuit pressure closes the valve, preventing backflow into the cabin. The open and egress positions provide manual override of the automatic position.

CO_2 Canister Selector Valve and CO_2 and Odor Removal Canisters. The CO_2 canister selector valve is a dual-flapper-type valve that routes flow through the CO_2 and odor removal canisters. The valve has primary and secondary positions. One flapper is at the inlet to the canisters: the other, at the outlet.

Each canister contains a cartridge filled with LiOH and activated charcoal. The primary canister cover has a debris trap, which may be replaced before, but not during, flight. A relief valve in the cartridge permits flow to bypass the debris trap if it becomes clogged. The canister selector valve is sufficiently leakproof to permit replacement of cartridges, with the cabin unpressurized.

Suit Circuit Fans. The suit circuit fans maintain the circulation of conditioned oxygen in the suit circuit. Each suit circuit fan is operated by a 28-volt d-c brushless motor, and each fan moves approximately 24 cfm at 25,000±500 rpm. Fan operation is controlled by the suit fan selector switch.

Suit Circuit Sublimator. The sublimator rejects suit circuit heat to space if the suit circuit heat exchanger is inoperative. It has a water inlet and a stream outlet that is vented overboard. Water and suit circuit gas both make a single pass through the unit, which comprises a stack of modules of several layers of porous plates, water, steam, and suit circuit gas passages.

Suit Circuit Heat Exchanger. The suit circuit heat exchanger is a duct-shaped unit of aluminum plate-and-fin construction. It has a single pass for both the coolant and the suit circuit gas and is used to remove excess heat in the ARS. Heat is transferred to the HTS coolant supplied to the heat exchanger.

Water Separator Selector Valve. The water separator selector valve is a manually operated, flapper-type valve that enables selection of either of two water separators. The valve handle is pushed in for separator No. 1 and pulled for separator No. 2.

Water Separators. Two water separators are connected in parallel, but only one is used at a time. Saturated gas and free moisture fed into the separator come in contact with the inlet sensor blades, which direct the flow onto a rotor at the proper angle. Most of the entrained moisture collects on the rotating perforated plate, which centrifuges the water into a rotating water trough. A stationary pitot tube, picks up the removed water and discharges it to the WMS. The pitot pumping action creates a dynamic head of pressure sufficient to ensure positive flow from the water separator to the WMS. Water passing through the perforated plate collects on a rotating conical drum and is fed into the water trough. The oxygen flow drives the rotating parts of the separator. A water drain boss on each separator drains the cavity between the rotating drum and the outer shell of the unit. Plumbing attached to each drain boss carries water away from this area and dumps it into a surge (collection) tank outside the suit circuit.

Suit Circuit Regenerative Heat Exchanger. The suit circuit regenerative heat exchanger is of the aluminum plate-and-fin type. Heat from the circulating warm coolant is transferred to the oxygen, which makes a single pass through the unit while the coolant makes two passes.

Suit Temperature Control Valve. The suit temperature control valve is a manually operated diverter valve that controls coolant flow through the suit circuit regenerative heat exchanger. The valve can be throttled from no bypass (no heating) to maximum bypass (maximum heating).

Suit Isolation Valve. The suit isolation valve is a manually operated, two-position dual-ball valve. In the suit flow position, suit-circuit gas is directed through the valve into the PGA, and from the PGA back into the suit circuit. In the suit disconnect position, the valve keeps the gas in the suit circuit, by passing the PGA's and preventing flow in either direction between the suit circuit and PGA's.

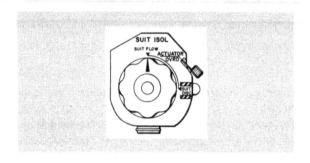

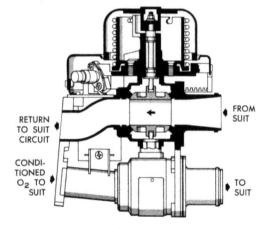

VALVE IN SUIT FLOW (OPEN) POSITION

R-38

Suit Isolation Valve

Setting the valve handle to suit flow loads a solenoid-operated spring return mechanism. A signal from the suit circuit pressure switch energizes the solenoid, releasing the return mechanism, which turns the valve to suit disconnect. The ACTUATOR OVRD lever enables manual release of the return mechanism to the SUIT DISC position. A valve position indicator switch provides a telemetry signal for SUIT DISC position.

Carbon Dioxide Partial Pressure Sensor. The carbon dioxide partial pressure sensor, in the suit circuit assembly, is a single-beam, dual-wavelength, filter photometer with ratio readout. The sensor operates on the infrared-absorption principle. It measures the amount of infrared energy absorbed by the carbon dioxide in a gas sample that passes through the sensor, by comparing transmitted energy of two different wavelengths in the infrared spectrum. (One wavelength is absorbed by carbon dioxide; the other is a reference.) This establishes an amplified ratio signal that is indicated as a d-c voltage proportional to the partial pressure of carbon dioxide in the gas sample.

The sensor has two sections: optics and electronics. The optics section has the infrared energy source (a small tungsten lamp), a collimating lens, a lens that reimages the source on the dual filter, an aperture to fix the source image on the dual filter, and a lens that reimages the chopped and filtered source image onto the detector target. The electronics section detects and decodes the signal, computes the ratio and, then, reads out a continuous d-c voltage proportional to the partial pressure of carbon dioxide in the gas sample. The sensor provides an electrical signal to the PART PRESS CO_2 indicator and a telemetry signal to indicate the carbon dioxide level in the gas supplied to the astronauts.

SUIT LIQUID COOLING ASSEMBLY

Cabin Fan. The fan motor is of the brushless, d-c type; it operates on 28 volts dc with an input power of 62 watts average, 210 watts peak. The fan circulates cabin gas and can move approximately 5 pounds of air per minute at 13,000 rpm. The fan permits operation at sea level for checkout purposes. An input voltage of 15 volts dc is provided for this purpose.

Water-Glycol Heat Exchanger. The water glycol heat exchanger transfers heat from the warm water returning from the LCG to the coolant of the heat transport section. This heat exchanger is of the cross-counterflow, single-pass water and multipass coolant type.

Liquid Garment Cooling Valve. The liquid garment cooling valve is a manually operated diverter valve that controls water flow to the water-glycol heat exchanger. Part or all of the water may be manually diverted around the heat exchanger to provide varying degrees of cooling, depending on astronaut needs.

Water Accumulator. The water accumulator consists of an aluminum housing, diaphragm, spring, diaphragm piston guide, and diaphragm piston. The system water pressure opposes the spring action in the accumulator to maintain the correct pressure level in the water loop. The accumulator which is filled, prior to launch, with 8.0 cubic inches of water, serves as a reservoir to make up for system leakage and volumetric changes due to temperature changes.

Water Pump. The water pump circulates cooling water through the suit liquid cooling assembly. The pump motor is of the oscillator type, with integral electronics to convert the DC supply to AC to drive the motor. A voltage regulator steps down the LM-supplied 28 volts dc to 16.5 volts dc for pump operation.

Suit Umbilical Water Hoses. The water umbilical hoses transport water between the LCG and the suit liquid cooling assembly. The hoses are flexible silicon rubber, covered with Beta cloth.

Multiple Water Connector. The multiple water connectors are quick disconnects that connect the water umbilical hoses to the LCG receptacle on the astronauts outer suit. The connectors provide dual flow into, and out of, the LCG. Poppet valves minimize leakage during connecting and disconnecting.

OXYGEN SUPPLY AND CABIN PRESSURE CONTROL SECTION

DESCENT OXYGEN TANKS

The descent oxygen tanks are in quad 3 and 4 of the descent stage. The tanks hold 48.01 pounds of oxygen, stored at a pressure of 2,690 psia. The tanks provides four PLSS refills at 1.60 pounds each and four cabin pressurizations at 5.5 pounds each. The tanks are filled through a high-pressure fill port, which is capped and lockwired before launch.

HIGH-PRESSURE OXYGEN CONTROL ASSEMBLY

The major components of the high-pressure oxygen control assembly are a high-pressure oxygen regulator, a bypass oxygen pressure relief valve, an overboard relief valve, and a burst diaphragm assembly.

High-Pressure Oxygen Regulator. The internally series redundant high-pressure oxygen regulator receives high-pressure oxygen from the descent oxygen tanks and regulates the pressure to 875 to 960 psia. If the upstream (primary) regulator sensor fails, the valve fails open, permitting the downstream (secondary) regulator to control outlet pressure. Descent oxygen that enters the regulator is sensed by the primary sensor. As the pressure builds up inside the sensor, it expands and allows the primary valve poppet to move towards its seat, regulating the outlet pressure to the secondary regulator. The secondary regulator operates the same as the primary regulator.

Bypass Oxygen Pressure Relief Valve. The bypass oxygen pressure relief valve protects the descent oxygen tanks against overpressurization by bypassing the high-pressure oxygen regulator. The valve is designed to fail in the open condition if it malfunctions. If the relief valve malfunctions (bellows ruptures), the valve is automatically placed in the failed-open condition, bypassing the oxygen to a secondary (identical) relief valve.

Overboard Relief Valve. The series-redundant overboard relief valve vents oxygen to ambient when

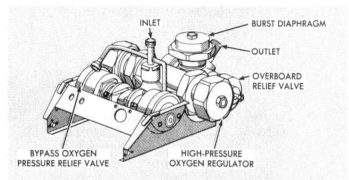

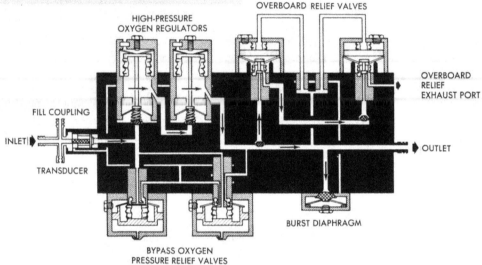

R-39

High Pressure Oxygen Control Assembly

the pressure downstream of the high-pressure oxygen regulator reaches 1,025 psig. The valve is fully open at 1,090 psig and reseats at 985 psig at +75° F. The relief valve has a fail-open feature. If a capsule sensing element leaks, the oxygen is dumped into the leakage chamber and bleeds overboard.

Burst Diaphragm Assembly. The burst diaphragm assembly opens when the flow from the descent oxygen tank exceeds the relieving capability of the overboard relief valve. An aluminum disk in the inlet port of the burst diaphragm assembly ruptures at a system pressure between 1,300 and 1,400 psig. System pressure causes the diaphragm support to move away from the disk, causing it to rupture. The disk support poppet opens and vents the descent oxygen tanks. The diaphragm assembly permits a minimum flow of approximately 10 pounds per minute. When descent tank pressure is

reduced to 1,000 psia, the disk support poppet reseats to maintain sufficient oxygen for one cabin repressurization up to 1 hour after disk rupture.

ASCENT OXYGEN TANKS

Two identical tanks supply all the oxygen required for metabolic consumption and to compensate for cabin and/or suit circuit leakage, oxygen component leakage, subsequent to switchover to ascent comsumables. Both tanks are in the aft equipment bay. Oxygen flow from either tank is controlled by individual oxygen shutoff valves on the oxygen control module.

OXYGEN CONTROL MODULE

The oxygen control module is mounted on the suit circuit package located to the right rear of the LM Pilot's station. It contains filters, a PLSS oxygen shutoff valve, descent and ascent oxygen shut-off valves, oxygen demand regulators, and a cabin repressurization and emergency oxygen valve. An interlock prevents opening the ascent oxygen shut-off valves until the descent oxygen shutoff valve is closed. Filters at the inlets to the oxygen control module and the PLSS disconnect remove particulate matter from the oxygen. Filtering capability is 18 microns nominal and 40 microns absolute.

PLSS and Oxygen Tank Shutoff Valves. Four positive-action, manually operated shutoff valves control oxygen flow into the oxygen control module. Rotating the valve handle to the open position displaces a spring-loaded seal. The valves have detents in the open and closed positions; mechanical stops on the handle prevent overtravel.

Oxygen Demand Regulators. Two, parallel oxygen demand regulators regulate oxygen from the tanks to the suit circuit. Both regulators are manually controlled by individual four-position selector handles. Suit circuit pressure is fed back to an aneroid bellows that actuates a poppet. Rotating the valve handle to cabin or egress position changes the spring force acting on the poppet and establishes two outlet pressure levels. Selecting the direct O_2 position, fully opens the valve. With an upstream pressure of 900 psia and a temperature of $+70°$ F, one regulator can flow approximately 7.0 pounds per hour into the suit circuit. In the closed position, an auxiliary poppet stops all flow through the regulator. Each regulator has valve position indicator (VPI) switches, and two functional switches that control electrical circuits in the ECS.

Cabin Repressurization and Emergency Oxygen Valve. The cabin repressurization and emergency oxygen valve is a solenoid-operated valve with manual override. It is used to repressurize the cabin after a deliberate decompression and provides an emergency flow of oxygen if the cabin is punctured. If the cabin is punctured and the diameter of the hole does not exceed 0.5 inch, the valve can maintain cabin pressure at 3.5 psia for at least 2 minutes, allowing the astronauts to return to a closed-suit environment. The cabin repressurization

and emergency oxygen valve is controlled by a three-position handle on the oxygen control module.

When the automatic position is selected, valve operation is controlled by the solenoid, which is actuated by the cabin pressure switch. If cabin pressure drops below normal, the cabin pressure switch energizes the solenoid. This moves the valve yoke to open the inlet valve and permits oxygen from the oxygen manifold to flow through the valve into the cabin. If cabin pressure increases to normal, the cabin pressure switch deenergizes the solenoid and spring action closes the valve. When supplied by the descent GOX tanks, flow is 4 pounds per minute maximum; when supplied by the ascent tanks, 8 pounds per minute at 500 psia. When the manual position is selected, the handle shaft actuates a cam that moves the yoke to open the inlet valve and establish flow to the cabin. In the close position an auxiliary poppet is forced onto its seat, shutting off the flow.

CABIN RELIEF AND DUMP VALVES

A cabin relief and dump valve is installed in each hatch. The valve is a differential-pressure, servo-actuated poppet valve that prevents cabin over-pressurization and permits manual dumping of cabin pressure. Each valve is controlled with either of two handles: one, inside the cabin; the other, outside. The inside handles can be used to select three positions (dump, automatic, or closed). The outside handles have only a dump position.

Normally, the cabin relief and dump valves are set to the automatic position. In this position, the valves are operated by the servo valve. When the cabin-to-ambient pressure differential reaches approximately 5.4 psi, oxygen is vented overboard and cabin pressure is reduced to an acceptable value. When the pressure differential is 5.8 psi, one fully open dump valve can dump 11.1 pounds of oxygen per minute overboard. When the valve handle inside the cabin is set to the automatic position, the valve can be opened manually from outside.

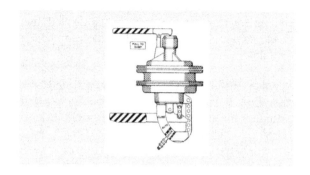

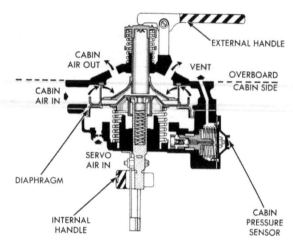

R-40

Cabin Relief and Dump Valve

Setting the internal handle to the dump position unseats the poppet. The valve can dump cabin pressure from 5.0 to 0.08 psia in 180 seconds without cabin oxygen inflow. Setting the handle to the closed position prevents the valve from opening at normal pressures, if the servo valve fails.

CABIN PRESSURE SWITCH

The cabin pressure switch monitors cabin pressure and enables electrical signals to control related ECS functions. The switch is an absolute-pressure device that consists of three separate hermetically sealed microswitch capsules. The capsules are set to cause switch closure when cabin pressure decays to 3.7 to 4.45 psia during pressurized-cabin operation.

If this occurs with either oxygen demand regulator in cabin mode, the suit gas diverter valve closes, the cabin repressurization and emergency oxygen valve opens, and the cabin warning light goes on. Increasing cabin pressure to 4.40 to 5.0 psia opens the cabin pressure switch circuits, closes the cabin repressurization and emergency oxygen valve, and deenergizes the cabin warning light.

WATER MANAGEMENT SECTION

DESCENT STAGE WATER TANKS

The descent stage water tanks are aluminum tanks with an internally mounted standpipe and bladder. The bladder contains the water; the space between the bladder is charged with nitrogen according to a schedule dependent on the load (48.2 psia maximum at $+80°$ F). The nitrogen forces the water out of the bladder through the standpipe and into the system. The tank outlets are connected to the water control module. Water from the descent water tanks is routed through the water tank select valve by setting it to the descent position.

Descent Water Valve. The descent water (DES H_2O) valve is a manually operated, poppet-type shutoff valve. The valve has open and closed positions. In the open position, the valve provides high-pressure water flow from the descent tanks to the water dispenser.

ASCENT STAGE WATER TANKS

The ascent stage water tanks are in the overhead unpressurized portion of the cabin. They are similar to the descent stage water tank, but are smaller. An initial nitrogen charge of 48.2 psia at $+80°$ F is used in each tank. The tank outlets are connected to the water control module. Water from the ascent water tanks is routed through the water tank selector valve by setting it to the ascent position.

Ascent Water Valve. The ascent water valve, at the top of the water control module, is a manually operated, poppet-type shutoff valve. The valve has open and closed positions. In the open position, the valve provides high-pressure water flow from the ascent tank to the water dispenser.

WATER CONTROL MODULE

The manual controls of the WMS are grouped together on the water control module. The module consists of check valves, shutoff valves, a water tank selector valve, and water pressure regulators. Each water tank outlet is connected to the module, which diverts the water to selected flow paths.

Check Valves. There are five check valves in the water control module: one in each tank feed line, and one in each discharge line from the ARS water separators. The check valves prevent water flow from the module to the water tanks and water separators.

Water Tank Selector Valve. The water tank selector valve is a manually operated, three-position, two-spool valve. The two spools (primary and secondary), linked to the valve handle, rearrange the internal ports to establish proper flow paths. The valve has descent, ascent, and secondary positions. In the descent position, the primary spool establishes a flow path between the descent water

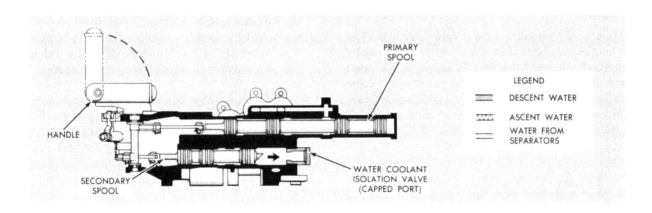

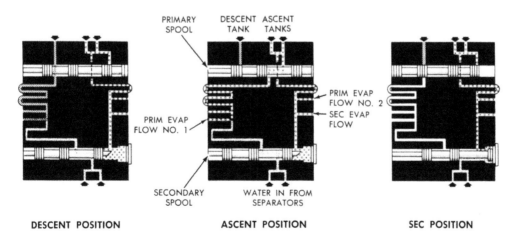

DESCENT POSITION · ASCENT POSITION · SEC POSITION

Water Tank Selector Valve

tanks and the primary water manifold. In the ascent position, the primary spool establishes a flow path between the ascent tanks and the primary water manifold. When the valve is in the secondary position, flow is diverted from the ARS separator to the secondary manifold and water is routed from the ascent water tanks to the secondary manifold.

Water Pressure Regulators. Four water pressure regulators are in the WMS. Two are in series in the primary manifold, one in the secondary manifold, and one is located external to the water module in the water pressure regulator module. The regulators contain a spring-loaded diaphragm that senses the differential between ARS reference pressure and the downstream water pressure. The diaphragm moves a balanced lever attached to a metering poppet. The water discharge pressure is maintained at 0.4 to 1.0 psi above the reference pressure.

Primary Evaporator Flow Valve No. 1. The primary evaporator flow No. 1 valve is a manually operated, poppet-type shutoff valve. It has open and closed positions. In the open position, the valve allows flow from the ascent or descent water tanks, through the primary regulators, to the primary sublimator.

Primary Evaporator Flow Valve No. 2. The primary evaporator flow No. 2 valve is a manually operated, poppet-type shutoff valve. It has open and closed positions. In the open position, the valve acts as a backup to the primary evaporator flow valve to provide ascent tank water from the secondary water manifold to the primary sublimator.

Secondary Evaporator Flow Valve. The secondary evaporator flow valve is a manually operated, poppet-type shutoff valve. It has open and close positions. The valve controls water flow from the secondary water manifold to the secondary sublimator and to the suit circuit sublimator.

Water Pressure Regulator Module. The water pressure regulator module consists of a pressure regulator and a manifold. The module is located in the secondary water circuit, downstream and in series with the secondary water pressure regulator in the water control module.

HEAT TRANSPORT SECTION

COOLANT ACCUMULATOR

There are two accumulators, one in the primary loop and an identical one in the secondary loop. The coolant accumulator consists of a two-piece aluminum cylinder that contains a movable spring-loaded piston bonded to a diaphragm. The fluid side contains approximately 46 cubic inches of fluid under pressure. The pressure varies, directly with fluid level, from 5.6 psia at 5% level to 8.0 psia at 80% level. The accumulator maintains a head of pressure on the glycol pump inlets to prevent pump cavitation and replaces coolant lost through subsystem leakage. The piston moves in response to volumetric changes caused by temperature variations in the primary loop or by leakage. The accumulator spring side is vented to space.

The accumulator has a low-level sensor with a switch that trips when only 10 ± 5% of the coolant volume remains. The switch is mounted on top of the accumulator. A tube extending from the top of the accumulator, houses a rod attached to the piston. The switch provides a telemetry signal.

PRIMARY COOLANT FILTER

The primary coolant filter has a filtering capability of 35 microns. It has an integral pressure relief bypass feature that opens at 0.27 to 0.4 psid to maintain coolant flow to the pumps if the filter becomes clogged.

COOLANT PUMP

The three coolant pumps are identical. Two pumps are connected in the primary loop; the third pump, in the secondary loop. The pumps are of the sliding-vane, positive-displacement type; they are driven by 28-volt d-c brushless motors. The motors are of wet or submerged design and are cooled by the recirculating coolant. The pump speeds will vary with changes in coolant temperatures and pump input voltages. As a result, system coolant flow rates may vary from 4.2 to 5.4 pounds per hour causing primary pump pressure rise of 12 to 25 psi and secondary pump pressure rise of 8 to 18 psi.

SUIT TEMPERATURE CONTROL VALVE

The suit temperature control valve is a manually operated diverter valve. The amount of warm coolant flowing through the suit circuit regenerative heat exchanger is regulated by the handle, which can be turned from hot to cold.

SUBLIMATORS

The porous-plate-type sublimators (one in the primary loop and another in the secondary loop) are identical, except that the primary sublimator has a larger capacity. Each sublimator has a coolant inlet and outlet, a water inlet, and a steam outlet. Water makes one pass through the unit; coolant makes six passes through the primary sublimator and four passes through the secondary sublimator. For proper sublimator operation, water pressure must exceed 4.0 psia, but be less than 6.5 psia. The water pressure must also be less than the suit circuit static pressure plus the head pressure from the water separators to the sublimator.

The unit rejects heat to space by sublimation of ice. Water from the WMS flows through the water passages, into the porous plates, and is exposed to space environment. The vacuum pressure is below the triple point of water; this causes an ice layer to develop within the pores and on the inner surface of the plates. As the hot coolant flows through the sublimator passages, heat transfers from the coolant to the water and to the ice layer. The ice sublimates from the solid state to steam without passing through a liquid state, rejecting its heat load overboard through a duct. The thickness of the ice layer varies with the heat load imposed on the sublimator, resulting in a regulated output temperature over a range of input temperatures.

COLD PLATES AND RAILS

Electronic equipment that requires active temperature control is cooled by cold plates and cold rails. Most flat cold plates are installed between electronic equipment and the LM structure in a manner that minimizes heat transfer from the structure to the coolant, to avoid a reduction of the coolant cooling capacity. The surrounding structure and equipment may have a temperature range of $0°$ to $+160°$ F. The remaining flat cold plates are installed directly on the electronic equipment without making contact with the LM structure. Cold rails are also structural members and are used in the aft equipment bay in the descent stage for the DSEA. The IMU has an integral cooling circuit. Cold plates and cold rails for equipment essential for mission abort have two independent coolant passages, one for the primary loop and one for the secondary loop.

The flat cold plates are brazed assemblies with inlet and outlet fittings. The coolant flows between two parallel sheets, which are connected by fins for increased heat transfer and structural strength. The internal fin arrangement of these cold plates ensures sufficient flow distribution.

The cold rails are channel-and-tube-type extrusions; the tubular part forms the inside center of the channel. The tube has fins and, at the ends, coolant inlet and outlet fittings. The cold rails are installed in parallel arrangement, with equal space between the rails to accommodate equipment designed for mounting on cold rails. Each cold rail (except the first and the last one) cools two adjacent rows of equipment.

CONTROLS AND DISPLAYS

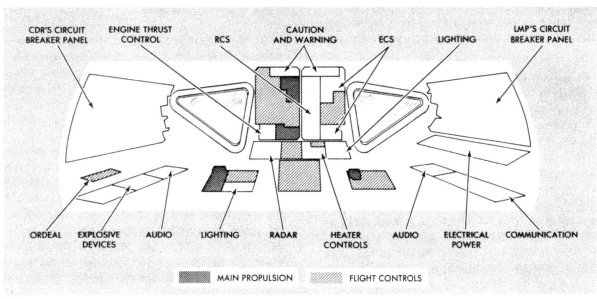

Control and Display Panels Locator

The controls and displays enable astronauts to monitor and manage the LM subsystems and to control the LM manually during separation, docking, and landing.

In general, the controls and displays are in subsystems groupings located in accordance with astronaut responsibilities. Certain controls and displays are duplicated to satisfy mission and/or safety requirements; a system of interlocks prevents simultaneous operation of these controls Controls and displays that enable either astronaut to control the LM are centrally located; these are accessible from both flight stations. Controls that could be operated inadvertently are appropriately guarded.

Annunciator displays go on if malfunctions occur in the LM subsystems; at the same time, two flashing master alarm lights and an alarm tone (in the astronaut headsets) are activated. Digital and analog displays provide the astronauts with subsystem-status information such as gas and liquid quantities, pressures, temperatures, and voltages.

There are 12 control and display panels. The main control and display panels (1 and 2) are canted and centered between the flight stations. Panels 3 and 4 are below these panels, within convenient reach and scan of both astronauts. Panels 5, 8, and 11 are located for use by the Commander. Panels 6, 12, 14, and 16 are located for use by the LM pilot.

Panel 1, directly in view of the Commander, contains various controls and displays, including warning lights, digital counters, navigational instruments, engine thrust control switches, and engine, fuel, and altitude indicators. Panel 2, directly in view of the LM Pilot, contains caution lights, reaction control indicators, environmental control indicators, navigational instruments, and various other indicators and switches. Panel 3 contains controls and displays for radar, stabilization and control, heater control, an event timer, and lighting. Panel 4 contains a display and keyboard assembly. The display and keyboard provides a two-way communications link between the astronauts and the LM guidance computer. The panel contains indicator lights, pushbuttons, data displays, and toggle switches. In front of the Commander's and LM Pilot's stations, at waist height, are panels 5 and 6, respectively. Panel 5 contains lighting and mission timer controls, engine start and stop pushbuttons, and an X-translation pushbutton. Abort guidance controls are on panel 6.

At the left of the Commander's station is panel 8, which is canted up 15° from the horizontal. This panel contains controls and displays for explosive devices and descent propulsion, and audio controls. The orbital rate display — earth and lunar (ORDEAL) panel aft and on top of panel 8, is an electromechanical device that provides an alternative to the pitch display of the flight director attitude indicator on panels 1 and 2. When selected, the ORDEAL produces a flight director attitude indicator display of computer local vertical attitude indicator display of computed local vertical attitude during earth or lunar circular orbits. Panel 11, directly above panel 8, has five angled surfaces that contain circuit breakers. Each row of circuit breakers is canted 15° to the line of sight, so that a white band around the circuit breakers is visible when the breakers are open.

At the right of the LM Pilot's station is panel 12, which is canted up 15° from the horizontal. This panel contains audio, communications, and communications antenna controls and displays. Directly above panel 12 is panel 14. It is canted up 36.5° from the horizontal and contains controls and displays for electrical power distribution and monitoring. Panel 16, directly above panel 14, has four angled surfaces that contain circuit breakers.

This makes a white band visible around the circuit breakers when the breakers are open.

At the right of each flight station is a pistol-grip control (attitude controller assembly), used to control LM attitude changes.

At the left of each flight station is a T-handle control (thrust/translation controller assembly). This assembly, an integrated translation and thrust controller, is used to command vehicle translations by firing thrusters in the Reaction Control Subsystem, and to throttle the descent engine.

An alignment optical telescope is located between and above the flight stations; it is a manually operated, unity-power, periscope-type device. It is used to determine the position of the LM, using a catalog of stars stored in the LM guidance computer and celestial measurements made by the astronauts. A utility light control centered above the flight stations comprises two switches and two portable light fixtures, one for the Commander and one for the LM Pilot.

Environmental Control Subsystem controls are grouped together on the aft bulkhead, behind the LM Pilot. These controls include an oxygen control module, a suit gas diverter, suit flow controls, suit circuit and canister controls, cabin/suit temperature controls, and water management controls. Cabin relief and dump valve controls are located on the forward and overhead hatch. The Environmental Control Subsystem controls enable the astronauts to maintain a habitable environment, decompress and repressurize the cabin, and regulate water flow for drinking, cooling, food preparation, and firefighting.

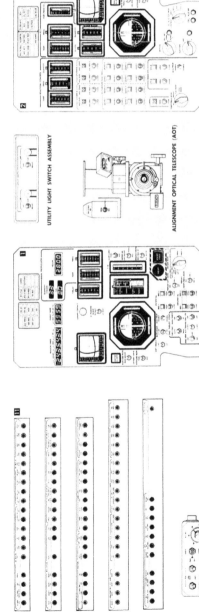

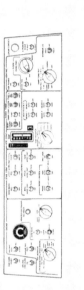
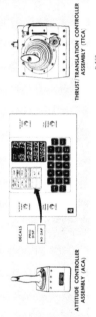
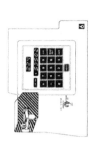

UTILITY LIGHT SWITCH ASSEMBLY

ALIGNMENT OPTICAL TELESCOPE (AOT)

ORBITAL RATE DISPLAY —
EARTH AND LUNAR (ORDEAL)

ATTITUDE CONTROLLER
ASSEMBLY (ACA)

THRUST / TRANSLATION CONTROLLER
ASSEMBLY (TTCA)

ATTITUDE CONTROLLER
ASSEMBLY (ACA)

THRUST / TRANSLATION CONTROLLER
ASSEMBLY (TTCA)

DECALS

Lunar Module Controls and Displays

R-43A

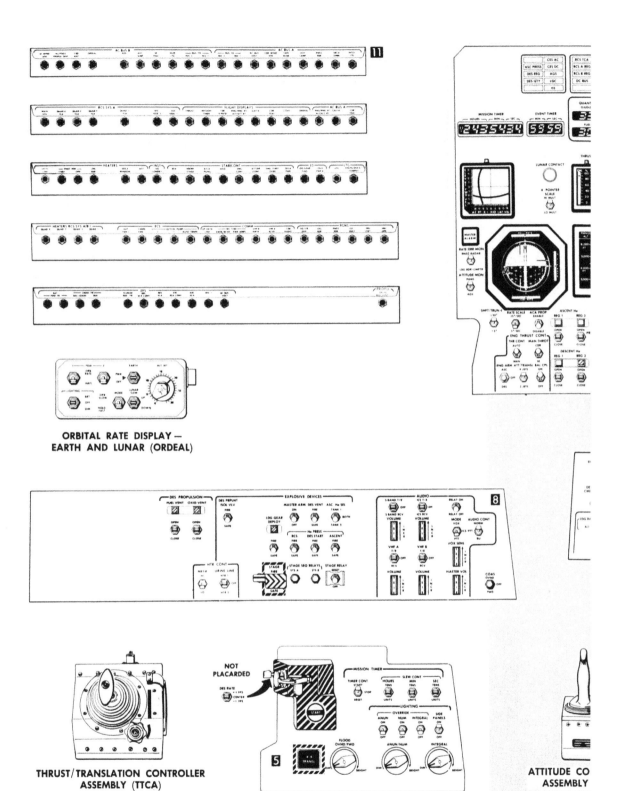

ORBITAL RATE DISPLAY —
EARTH AND LUNAR (ORDEAL)

THRUST/TRANSLATION CONTROLLER
ASSEMBLY (TTCA)

ATTITUDE CO
ASSEMBLY

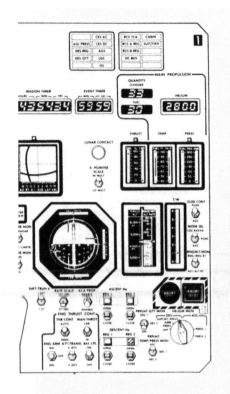

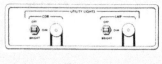

UTILITY LIGHT SWITCH ASSEMBLY

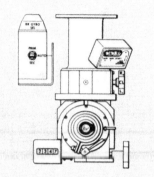

ALIGNMENT OPTICAL TELESCOPE (AOT)

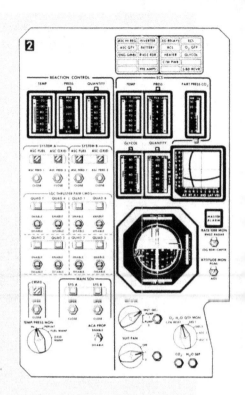

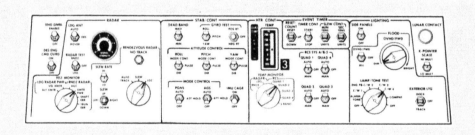

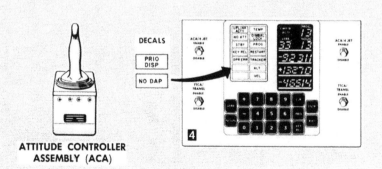

DECALS

PRIO DISP

NO DAP

ATTITUDE CONTROLLER ASSEMBLY (ACA)

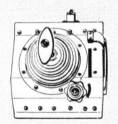

THRUST/TRANSLATION CONTROLLER ASSEMBLY (TTCA)

R-43A

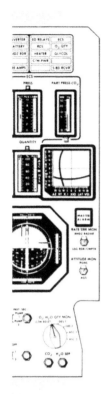

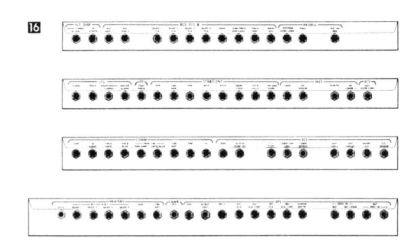

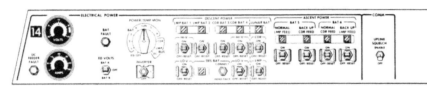

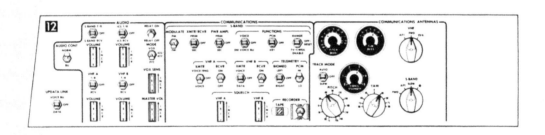

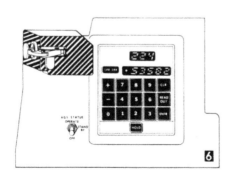

NTROLLER

**ATTITUDE CONTROLLER
ASSEMBLY (ACA)**

Lunar Module Controls and Displays

GUIDANCE, NAVIGATION, AND CONTROL

QUICK REFERENCE DATA

PRIMARY GUIDANCE AND NAVIGATION SECTION

Navigation base
    Weight   3 pounds
    Diameter   14 inches
    Leg length (approx)   10 inches
    Material   Aluminum

Inertial measurement unit
    Weight (approx)   42 pounds
    Diameter   12.5 inches
    Temperature   +126° F

Alignment optical telescope
    Number of detent positions   6
    Field of view of each detent   60°
    Counter readout   000.00° to 359.98°
    Length   36 inches

Computer control and reticle dimmer assembly
    Height   3-3/8 inches
    Width   4-3/8 inches
    Depth   2-1/2 inches
    Weight   3 pounds

Pulse torque assembly
    Height   2-1/2 inches
    Width   11 inches
    Depth   13 inches
    Weight   15 pounds

Power and servo assembly
    Height   2-5/8 inches
    Width   8-7/8 inches
    Depth   23-1/2 inches
    Weight   20 pounds

Coupling data unit
    Number of channels   5
    Height   5-1/2 inches
    Width   11-1/3 inches
    Depth   20 inches
    Weight   35 pounds

LM guidance computer
    Computer type   Automatic, electronic, digital, general-purpose and control
    Internal transfer   Parallel (all bits simultaneously)
    Memory   Random access
        Erasable   Coincident-current core; 2,048-word capacity
        Fixed   Core rope; 36,864-word capacity

LM guidance computer (cont)

Height	6 inches
Width	12-1/2 inches
Depth	24 inches
Weight	70 pounds
Word length	16 bits
Number system	Binary 1's complement - for manipulation
Circuitry type	Flat pack, NOR micrologic
Memory cycle time	12 microseconds
Add time	24 microseconds
Basic clock oscillator	2.048 mHz
Power Supplies	One +4-volt and one +14-volt switching regulator; operated from 28-volt d-c input power
Logic	Positive (Positive dc = Binary 1; 0 volts = Binary 0)
Parity	Odd

Display and keyboard

Height	8 inches
Width	8 inches
Depth	7 inches
Weight	17 pounds

Signal conditioner assembly

Height	3.3 inches
Width	8.8 inches
Depth	11.2 inches
Weight	5.6 pounds

ABORT GUIDANCE SECTION

Data entry and display assembly

Height	7.3 inches
Width	6.6 inches
Depth	5.6 inches
Weight	8.4 pounds
Logic levels	Zero: 0 to 0.5 vdc
	One: +3 to +5 vdc
Clock frequency	128 kpps

Abort electronics assembly

Computer type	Automatic, electronic, digital, general-purpose
Height	23.7 inches
Width	9.0 inches
Depth	5.0 inches
Weight	32.5 pounds
Power	12.5 watts (standby)
	96.0 watts (operate)

Abort electronics assembly (cont)

Logic levels	Zero: 0 to 0.5 vdc
	One: +3 to +5 vdc
Clock frequency	1,024 kpps
Memory capacity	4,096 words
Fixed	2,048 words
Erasable	2,048 words
Word size	18 bits

Abort sensor assembly

Height	5.1 inches
Width	9.0 inches
Depth	13.5 inches
Weight	20.7 pounds (with support)
Clock frequency	128 kpps
Operating temperature	$+120^\circ$ F

CONTROL ELECTRONICS SECTION

Attitude and translation controller assembly

Input signals	Attitude error, command rate, and rate gyro output
Operating frequency	800 Hz
Cooling	Conduction through mounting flanges
Temperature range	0° to $+160^\circ$ F

Rate gyro assembly

Input power	Single- and three-phase, 800 Hz
Starting power	1.8 watts (maximum; three-phase)
Input range	-25° to $+25^\circ$ per second
Input rate frequency	20 ± 4 Hz

Descent engine control assembly

A-C input power nominal voltage	115 vrms
Operating temperature range	$+57^\circ$ to $+97^\circ$ F
D-C input power nominal voltage	+4, +15, +28, and -15 volts dc
Total power consumption	7.9 watts (maximum)

Gimbal drive actuator

A-C power	115 ± 2.5 vrms, single phase, 400 Hz
A-C power consumption (steady-state average)	35 watts
Stroke	+2 to -2 inches $\pm 5\%$
Gimbal position	$+6^\circ$ to $-6^\circ \pm 5\%$
Gimbal rate	0.2°/sec $\pm 10\%$
Frequency of operation	5.0 Hz (maximum)

Attitude controller assembly

Operating power	28 volts. 800 Hz
Type of sensor	Proportional transducer
Displacement	0.28 volt/degree

Thrust/translation controller assembly

Operating power	28 volts, 800 Hz
Type of sensor	Proportional transducer

LANDING RADAR

Velocity sensor	Continuous-wave, three-beam
Radar altimeter	Frequency modulated/continuous wave (FM/CW)
Altitude capability	10 to 40,000 feet
Velocity capability	From altitude of 24,000 feet
Weight (approx)	43.5 pounds
Power consumption	125 watts dc (nominal)
	147 watts dc (maximum)
Heater power consumption	63 watts dc (maximum)
Altimeter antenna	
Type	Planar array, space-duplexed
RF power	100 mw (minimum)
Velocity sensor antenna	
Type	Planar array, space-duplexed
RF power	200 mw (minimum)
Transmitter frequency	
Velocity sensor	10.51 gHz
Radar altimeter	9.58 gHz
Warmup time	1 minute
FM sweep duration	0.007 second
Acquisition time	12 seconds (maximum)
Primary power	25 to 31.5 volts dc (nominal)
	3.5 to 6.5 amperes
Temperature range	
Electronics assembly	-20° to $+110^\circ$ F
Antenna assembly	$+50^\circ$ to $+150^\circ$ F

RENDEZVOUS RADAR AND TRANSPONDER

Rendezvous radar

Radar radiation frequency	9832.8 mHz
Radar received frequency	9792.0 mHz ± Doppler
Radiated power	300 mw (nominal)
Antenna design	Cassegrainian
Angle-tracking method	Amplitude monopulse
Antenna diameter	24 inches
Antenna beamwidth	4.0°
Gyroscopes	4 (two redundant)
Modulation	Phase modulation by three tones: 200 Hz, 6.4 kHz, and 204.8 kHz
Receiver channels	Reference, shaft (pitch), and trunnion (yaw)
Receiver intermediate frequencies	40.8, 6.8, and 1.7 mHz
Range	80 feet to 400 nm
Range accuracy	1% or 80 feet for ranges between 80 feet and 5 nm; or 300 feet for ranges between 5 and 400 nm
Range data output	15-bit serial format
Range rate	+4,900 to –4,900 fps
Range rate accuracy	±1 fps
Complete acquisition time	15 seconds

Angular accuracy

5 to 400 nm	0.12° to 0.24°

Transponder

Weight	16.0 pounds
Antenna	4-inch Y-horn, linearly polarized 12-inch interconnecting waveguide
Transmit frequency	9792 mHz
Receive frequency	9832.8 mHz ± one-way Doppler
Radiated power	300 mw
Acquisition time	1.8 seconds with 98% probability
Intermediate frequencies	
First IF	40.8 mHz
Second IF	6.8 mHz
Modulation	Phase modulation by three tones (200 Hz, 6.4 kHz, 204.8 kHz)
Range	80 feet to 400 nm
Range accuracy	Equal to maximum ranging error
Range rate accuracy	0.25% or 1 fps (whichever is greater)
Input power	75 watts
Heater	20 watts (maximum)

The LM is designed to take two astronauts from the orbiting CSM to the lunar surface and back again. The primary function of the Guidance, Navigation, and Control Subsystem (GN&CS) is accumulation, analysis, and processing of data to ensure the LM follows a predetermined flight plan at all times. To perform these functions, the guidance portion must know present position and velocity with respect to the guidance goal. The GN&CS provides navigation, guidance, and flight control to accomplish the specific guidance goal.

The astronaut is an active and controlling element of the LM. He can monitor information to and from the various LM subsystems and can manually duplicate the various control functions. During completely automatic flight, the astronaut functions as a monitor and decision maker; during semiautomatic flight, he is a controlling influence on the automatic system; and during manually controlled flight he may perform all GN&CS functions himself. The astronaut can also initiate an optical sighting program, utilizing celestial objects to align the guidance equipment.

Using cabin displays and controls, the astronaut can select modes of operation necessary to perform a desired function. In some mission phases, sequencing of modes of operation is automatically controlled by a computer. As calculations are performed by the computer, the results are displayed for astronaut evaluation and verification with ground-calculated data.

In the event of failure of automatic control, the astronaut manually controls the LM and performs vehicle flight control normally performed by the computer. He does this with a pair of hand controllers, which control attitude and translation, and with other controls on the cabin panels.

For purposes of the following discussion, a distinction is made between guidance (orbital alteration or redirection of the LM) and navigation (accumulation and processing of data to define the proper guidance to be accomplished).

NAVIGATION AND THE LUNAR MODULE

LM navigation involves the determination of the vehicle's present position and velocity so that the guidance function can plot the trajectory that the LM must follow.

When flying an aircraft between two points on earth, both points remain fixed with respect to each other. In spaceflight, however, the origin of the spacecraft's path and its destination or target are moving rapidly with respect to each other.

To determine the present position of the LM, celestial navigation is used to align the guidance system. This is accomplished by determining the vehicle's position in relation to certain fixed stars. Even though the stars may be moving, the distance that they move in relation to the total distance of the stars from the vehicle is so small that the stars can be thought of as being stationary.

The optical device which the astronauts use for navigation is an alignment optical telescope (AOT) protruding through the top of the vehicle and functioning as a sextant. The astronauts use it to take direct visual sightings and precise angular measurements of pairs of celestial objects. These measurements are transferred by the astronaut to the guidance elements to compute the position of the vehicle and to perform alignment of an inertial guidance system. There is a direct relationship between the angular measurements taken with the telescope and the mounting position of the telescope. The computer program knows the telescope's mounting position which is in alignment with the LM body axes and from this knowledge and astronaut-generated information, the computer is able to calculate the LM position.

During the landing phase and subsequent rendezvous phase, the LM uses radar navigational techniques to determine distance and velocity. Each phase uses a radar designed specifically for that phase (rendezvous radar, landing radar). Both radars inform the astronaut and the computer concerning position and velocity relative to acquired target. During lunar landing, the target is the surface of the moon; during rendezvous, the target is the Command Module.

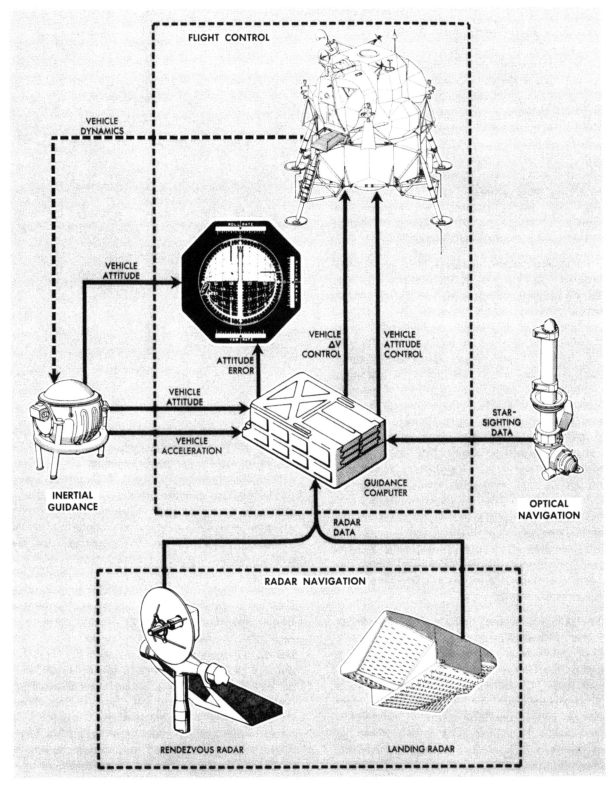

FLIGHT CONTROL

VEHICLE
DYNAMICS

VEHICLE
ATTITUDE

ROLL RATE

ATTITUDE
ERROR

VEHICLE
ATTITUDE

VEHICLE
ACCELERATION

VEHICLE
ΔV
CONTROL

VEHICLE
ATTITUDE
CONTROL

STAR-
SIGHTING
DATA

**INERTIAL
GUIDANCE**

GUIDANCE
COMPUTER

**OPTICAL
NAVIGATION**

RADAR
DATA

RADAR NAVIGATION

RENDEZVOUS RADAR

LANDING RADAR

R-44A

Guidance, Navigation, and Flight Control Functions

GUIDANCE AND THE LUNAR MODULE

After the position and velocity of the LM are determined, the guidance function establishes the steering for the predetermined flight path. Since objects in space are moving targets (as compared to those on earth, which are stationary), the guidance problem involves aiming not at the target's present position but at the position in which it will be when the vehicle path intersects the target path. On earth, the guidance problem is a two-dimensional one; it involves only longitude and latitude. In space, a third dimension is introduced; position cannot be plotted in earth terms.

To calculate the guidance parameters, a reference coordinate frame must be determined. A three-axis, right-hand, orthogonal, coordinate frame (inertial reference frame) is used. It is fixed in space and has an unchanging angular relationship with the stars. Its dimensional axes are designated as X, Y, and Z, and all spacecraft positions and velocities are related to this frame. The astronaut establishes this frame by sighting of celestial objects using the AOT. The vertical axis is designated as the X-axis. Its positive direction is from the descent stage to the ascent stage, passing through the overhead hatch. The lateral axis is designated as the Y-axis. Its positive direction is from left to right across the astronauts shoulders when they are facing the windows in the LM cabin. To complete the three-axis orthogonal system, the Z-axis is perpendicular to the X and Y axes. This axis is referred to as the forward axis, because +Z-axis direction is through the forward hatch. The +Z-axis is also used as the zero reference line for all angular measurements.

The guidance system based on this coordinate frame is referred to as an inertial guidance system. Inertial guidance provides information about the actual path of the vehicle in relation to a predetermined path. All deviations are transmitted to a flight control system. The inertial guidance system performs these functions without information from outside the vehicle. The system stores the predetermined flight plan, then automatically but not continuously, computes distance and velocity for a given mission time (called the state vector) of the vehicle to compensate, through vehicle control, for changes in direction.

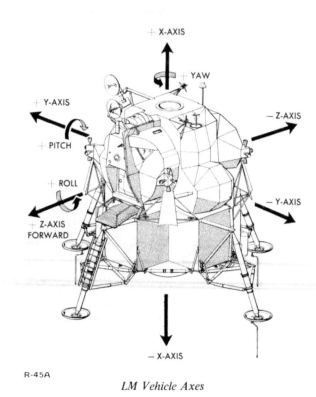

R-45A

LM Vehicle Axes

Inertial guidance systems are based on measurements made by accelerometers mounted on a structure called the stable member or platform. The stable member, in turn, is mounted inside three spherical gimbals, one for each principal axis of motion. Gyroscopes mounted on the stable member drive the gimbals to isolate the stable member from changes in LM attitude and hold the stable member in a fixed inertial position.

During flight, the stable member's axes must be held in fixed relation to the inertial reference frame regardless of the LM motion; otherwise resolvers mounted on each gimbal issue error signals. These error signals are used by the computer to generate commands to correct the attitude of the LM. The rotational axes of the LM are designated as yaw, pitch, and roll. Yaw rotation, about the X-axis affects the vehicle in the Y-Z plane. The effect is analogous to spinning around one's heels. Pitch rotation, about the Y-axis, affects the vehicle in the X-Z plane. The effect is analogous to a gymnast performing a somersault. Roll rotation, about the Z-axis, affects the vehicle in the X-Y plane. The effect is analogous to a person doing a cartwheel. Positive rotation is determined by the right-hand rule. This involves placing the thumb of the right hand in the positive direction of the axis about which rotation is to be determined. Then the remainder of the fingers are curled around the axis. The direction in which the fingers point is considered the direction of positive rotation.

FLIGHT CONTROL AND THE LUNAR MODULE

Flight control involves controlling the LM trajectory (flight path) and attitude. Flight path control depends on the motion of the LM center of gravity; attitude control primarily involves rotations about the center of gravity.

In controlling the LM in its flight path, the thrust of its engines must be directed so that it produces a desired variation in either magnitude or direction to place the LM in some particular orbit, position, or attitude. The major velocity changes associated with the lunar orbit, injection, landing, and ascent phases of the mission are accomplished by either the descent propulsion section or ascent

propulsion section of the Main Propulsion Subsystem (MPS). The engines can produce high thrust in specific directions in inertial space.

During the descent phase, the LM must be slowed (braked) to place it in a transfer orbit from which it can make a soft landing on the lunar surface. To accomplish braking, descent engine thrust is controllable so that the precise velocity (feet per second) necessary to alter the vehicle's trajectory can be achieved. For a soft landing on the lunar surface, the weight of the LM must be matched by an upward force so that a state of equilibrium exists, and from this point, the descent engine is shut off and the LM free falls to the lunar surface. The thrust of the descent engine provides this upward force, and since the weight of the vehicle is a variable (due to consumption of expendables) this is another reason why the magnitude of the engine thrust is controllable. In addition, the center of gravity is also variable and the thrust must be such that it is in line with the LM center of gravity. This is accomplished by gimbaling (tilting) the descent engine.

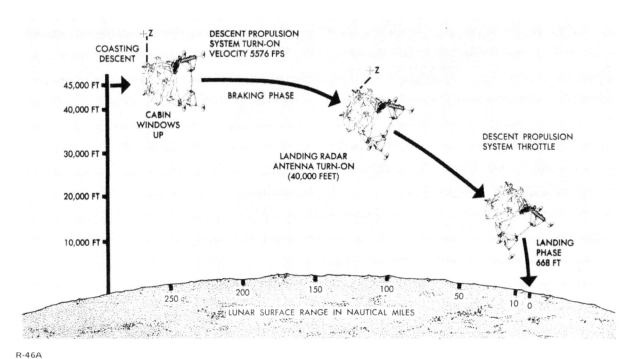

LM Powered Descent Profile

R-46A

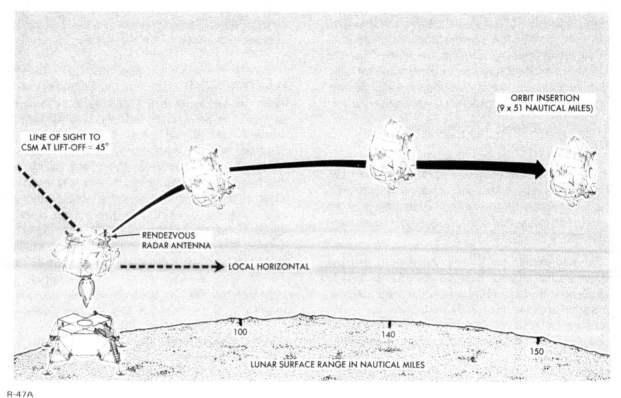

R-47A

LM Powered Ascent Profile

During the lunar ascent phase, the flight control portion of the GN&CS commands the ascent engine. In this phase, control of the thrust direction is not achieved by gimbaling the engine, but by attitude control, using the Reaction Control Subsystem (RCS) thrusters. This is necessary during ascent to keep the vehicle stabilized, because the center of gravity changes due to propellant depletion. The ascent engine is not throttleable, since the function of this engine is to lift the ascent stage from the lunar surface and conduct rendezvous. The proper orbit for rendezvous is achieved by means of a midcourse correction (if necessary) in which thrust is directed by attitude control, and thrust magnitude is controlled by controlling the duration of the burn.

It is apparent then for flight control, that some measure of the LM velocity vector and its position must be determined at all times for purposes of comparison with a desired (predetermined) velocity vector, at any particular instant, to generate an error signal if the two are not equal. The flight control portion of the primary guidance and navigation section then directs the thrust to reduce the error to zero.

Attitude control maintains the LM body axes in a fixed relationship to the inertial reference axes. Any pitch, roll, or yaw rotations of the vehicle produce a misalignment between the LM axes and where the LM axes should be. This is called attitude error and is detected by the inertial guidance system, which, in turn, routes the errors to the computer. The computer generates on and off commands for the RCS to reduce the error to zero. Attitude control is implemented through 16 rocket engine thrusters (100-pounds thrust each) equally distributed in clusters of four around the ascent stage. Each cluster is located so that it will exert sufficient torque to rotate the LM about its center of gravity. The thrusters are capable of repeated starts and very short (fraction of second) firing times. The appropriate thrusters are selected by the computer during automatic operation and manually by the astronaut during manual operation.

GUIDANCE, NAVIGATION, AND CONTROL SUBSYSTEM

To accomplish guidance, navigation, and control, the astronauts use 55 switches, 45 circuit breakers, and 13 indicators which interface with the various GN&CS equipment. This equipment is functionally contained in a primary guidance and navigation section, an abort guidance section, a control electronics section, and in the landing and rendezvous radars.

The primary guidance and navigation section (PGNS) provides, as the name implies, the primary means for implementing inertial guidance and optical navigation for the LM. When aided by either the rendezvous radar or the landing radar, the section provides for radar navigation. The section when used in conjunction with the control electronics section (CES) provides automatic flight control. The astronauts can supplement or override automatic control, with manual inputs.

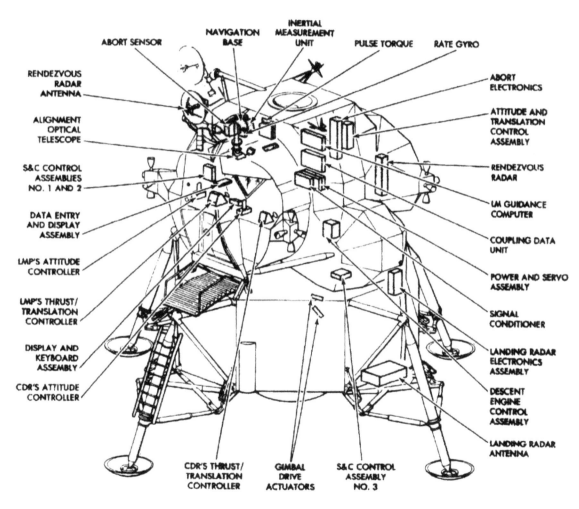

R-48

Guidance, Navigation, and Control Major Equipment Location

The abort guidance section (AGS) is primarily used only if the primary guidance and navigation section malfunctions. If the primary guidance and navigation section is functioning properly when a mission is aborted, it is used to control the LM. Should the primary section fail, the lunar mission would have to be aborted; thus, the term "abort guidance section." Abort guidance provides only guidance to place the LM in a rendezvous trajectory with the CSM or in a parking orbit for CSM-active rendezvous. The navigation function is performed by the primary section, but the navigation information also is supplied to the abort section. In case of a primary guidance malfunction, the abort guidance section uses the last navigation data provided to it. The astronaut can update the navigation data by manually inserting rendezvous radar data into the abort guidance section.

These integrated sections allow the astronauts to operate the LM in fully automatic, several semi-automatic, and manual control modes.

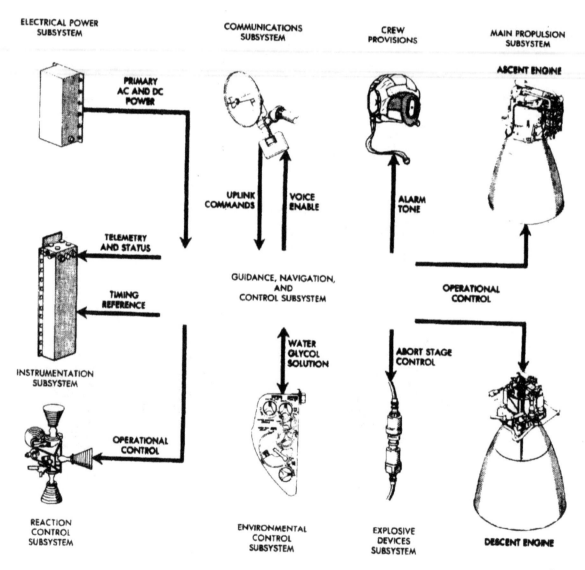

ELECTRICAL POWER SUBSYSTEM

COMMUNICATIONS SUBSYSTEM

CREW PROVISIONS

MAIN PROPULSION SUBSYSTEM

ASCENT ENGINE

PRIMARY AC AND DC POWER

UPLINK COMMANDS

VOICE ENABLE

ALARM TONE

TELEMETRY AND STATUS

TIMING REFERENCE

GUIDANCE, NAVIGATION, AND CONTROL SUBSYSTEM

OPERATIONAL CONTROL

INSTRUMENTATION SUBSYSTEM

WATER GLYCOL SOLUTION

ABORT STAGE CONTROL

OPERATIONAL CONTROL

REACTION CONTROL SUBSYSTEM

ENVIRONMENTAL CONTROL SUBSYSTEM

EXPLOSIVE DEVICES SUBSYSTEM

DESCENT ENGINE

R-49

GN&CS Relationship to Other Subsystems

Because the astronauts frequently become part of the control loop in this highly flexible system, a great deal of information must be displayed for their use. These displays include attitude and velocity, radar data, fuel and oxidizer parameters, caution and warning information, total velocity change information, timing and other information to assist them in completing their mission.

PRIMARY GUIDANCE AND NAVIGATION SECTION

The primary guidance and navigation section acts as an autopilot in controlling the LM throughout the mission. Normal guidance requirements include transferring the LM from a lunar orbit to its descent profile, achieving a successful landing at a preselected or crew-selected site, and performing a powered ascent maneuver which results in terminal rendezvous with the CSM. If the mission is to be aborted, the primary guidance and navigation section performs guidance maneuvers that place the LM in a parking orbit or in a trajectory that intercepts the CSM.

The navigational functional requirement of the section is that it provides the navigational data required for LM guidance. These data include line-of-sight (LOS) data from the AOT for inertial reference alignment, signals for initializing and aligning the abort guidance section, and data to the astronauts for determining the location of the computed landing site.

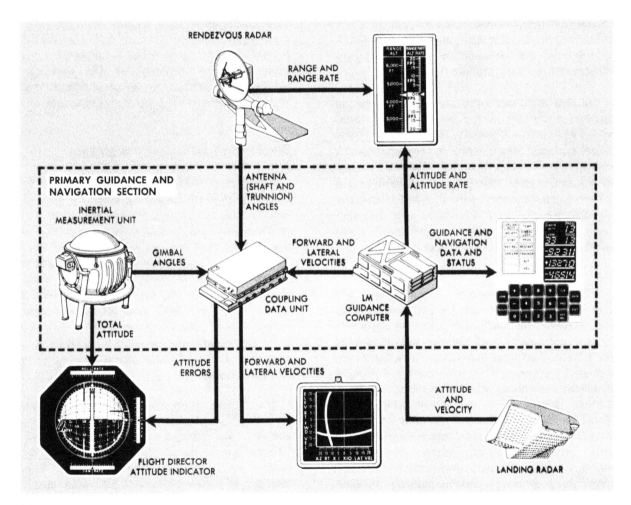

R-50A

Primary Guidance Data Displayed

The primary guidance and navigation section includes three major subsections: inertial, optical, and computer. Individually or in combination they perform all the functions mentioned previously.

The inertial subsection establishes the inertial reference frame that is used as the central coordinate system from which all measurements and computations are made. The inertial subsection measures attitude and incremental velocity changes, and assists in converting data for computer use, onboard display, or telemetry. Operation is started automatically by the guidance computer or by an astronaut using the computer keyboard. Once the subsection is energized and aligned to the inertial reference, any LM rotation (attitude change) is sensed by the stable member. All inertial measurements (velocity and attitude) are with respect to the stable member. These data are used by the computer in determining solutions to the guidance problems.

The optical subsection is used to determine the position of the LM, using a catalog of stars stored in the computer and celestial measurements made by an astronaut. The identity of celestial objects is determined before earth launch. The AOT is used by the astronaut to take direct visual sightings and precise angular measurements of a pair of celestial objects. The computer subsection uses this data along with prestored data to compute position and velocity and to align the inertial components.

The computer subsection, as the control and data processing center of the LM, performs all the guidance and navigation functions necessary for automatic control of the path and attitude of the vehicle. For these functions, the GN&CS uses a digital computer. The computer is a control computer with many of the features of a general-purpose computer. As a control computer, it aligns the stable member and positions both radar antennas. It also provides control commands to both radars, the ascent engine, the descent engine, the RCS thrusters, and the LM cabin displays. As a general-purpose computer, it solves guidance problems required for the mission.

ABORT GUIDANCE SECTION

The abort guidance section is used as backup for the primary guidance and navigation section during a LM mission abort. It determines the LM trajectory or trajectories required for rendezvous with the CSM and can guide the LM from any point in the mission, from LM-CSM separation to LM-CSM rendezvous and docking, including ascending from the lunar surface. It can provide data for altitude displays, for making explicit guidance computations and also issue commands for firing and shutting down engines. Guidance can be accomplished automatically or manually by the astronauts, based on data from the abort guidance section.

The abort guidance section is an inertial system rigidly strapped to the LM rather than mounted on a stabilized platform. Use of the strapped-down inertial system, rather than a gimbaled system, offers sufficient accuracy for LM missions, at savings in size and weight. Another feature is that it can be updated with radar and optical aids.

CONTROL ELECTRONICS SECTION

The control electronics section processes RCS and MPS control signals for vehicle stabilization and control. To stabilize the LM during all phases of the mission the control electronics section provides signals that fire any combination of the 16 RCS thrusters. These signals control attitude and translation about or along all axes. The attitude and translation control data inputs originate from the primary guidance and navigation section during normal automatic operation from two hand controllers during manual operations, or from the abort guidance section during certain abort situations.

The control electronics section also processes on and off commands for the ascent and descent engines, and routes automatic and manual throttle commands to the descent engine. Trim control of the gimbaled descent engine is also provided to assure that the thrust vector operates through the LM center of gravity.

LANDING RADAR

The landing radar, located in the descent stage, provides altitude and velocity data during lunar descent. The primary guidance and navigation section calculates control signals for descent rate, hovering, and soft landing. Slant range data begins at approximately 40,000 feet above the lunar surface; velocity data, at approximately 35,000 feet.

The landing radar uses four microwave beams; three to measure velocity by Doppler shift continuous wave, one to measure altitude by continuous-wave frequency modulation.

RENDEZVOUS RADAR

The rendezvous radar, operated in conjunction with a CSM transponder, acquires and tracks the CSM before and during rendezvous and docking. The radar, located in the ascent stage, tracks the CSM during the descent phase of the mission to supply tracking data for any required abort maneuver and during the ascent phase to supply data for rendezvous and docking. When the radar tracks the CSM, continuous measurements of range, range rate, angle, and angle rate (with respect to the LM) are provided simultaneously to the primary guidance and navigation section

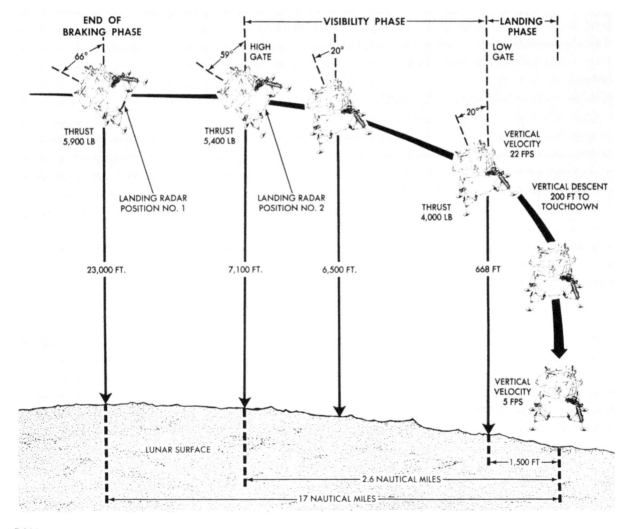

R-51A

Nominal Descent Trajectory from High Gate to Touchdown

and to LM cabin displays. This allows rendezvous to be performed automatically under computer control, or manually by the astronauts. During the rendezvous phase, rendezvous radar performance is evaluated by comparing radar range and range rate tracking values with MSFN tracking values.

The CSM transponder receives an X-band three-tone phase-modulated, continuous-wave signal from the rendezvous radar, offsets the signal by a specified amount, and then transmits a phase-coherent carrier frequency for acquisition by the radar. This return signal makes the CSM appear as the only object in the radar field of view. The transponder provides the long range (400 nm) required for the mission.

The transponder and the radar use solid-state varactor frequency-multiplier chains as transmitters, to provide high reliability. The radar antenna is space stabilized to negate the effect of LM motion on the line-of-sight angle. The gyros used for this purpose are rate-integrating types; in the manual mode they also supply accurate line-of-sight, angle-rate data for the astronauts. Range rate is determined by measuring the two-way Doppler frequency shift on the signal received from the transponder. Range is determined by measuring the time delay between the received and the transmitted three-tone phase-modulated waveform.

FUNCTIONAL DESCRIPTION

The GN&CS comprises two functional loops, each of which is an independant guidance and control path. The primary guidance path contains elements necessary to perform all the functions required to complete the LM mission. If a failure occurs in this path the abort guidance path can be substituted. To understand these two loops, the function of each major component of GN&CS equipment must be known.

PRIMARY GUIDANCE AND NAVIGATION SECTION

INERTIAL SUBSECTION

The inertial subsection consists of a navigation base, an inertial measurement unit, a coupling data unit, pulse torque assembly, power and servo assembly, and signal conditioner assembly.

The navigation base is a lightweight mount that supports, in accurate alignment, the inertial measurement unit (IMU), the AOT, and an abort sensor assembly (part of the abort guidance section). Structurally, it consists of a center ring with four legs that extend from either side of the ring. The inertial measurement unit is mounted to the legs on one end and the telescope and the abort sensor assembly are mounted on the opposite side.

The inertial measurement unit is the primary inertial sensing device of the LM. It is a three-degree-of-freedom, stabilized device that maintains an orthogonal, inertially referenced coordinate system for LM attitude control and maintains three accelerometers in the reference coordinate system for accurate measurement of velocity changes.

The coupling data unit converts and transfers angular information between the navigation and guidance hardware. The unit is an electronic device that performs analog-to-digital and digital-to-analog conversions. The coupling data unit processes the three attitude angles associated with the inertial reference and the two angles associated with the rendezvous radar antenna.

The pulse torque assembly supplies inputs to, and processes outputs from, the inertial components in the inertial subsection.

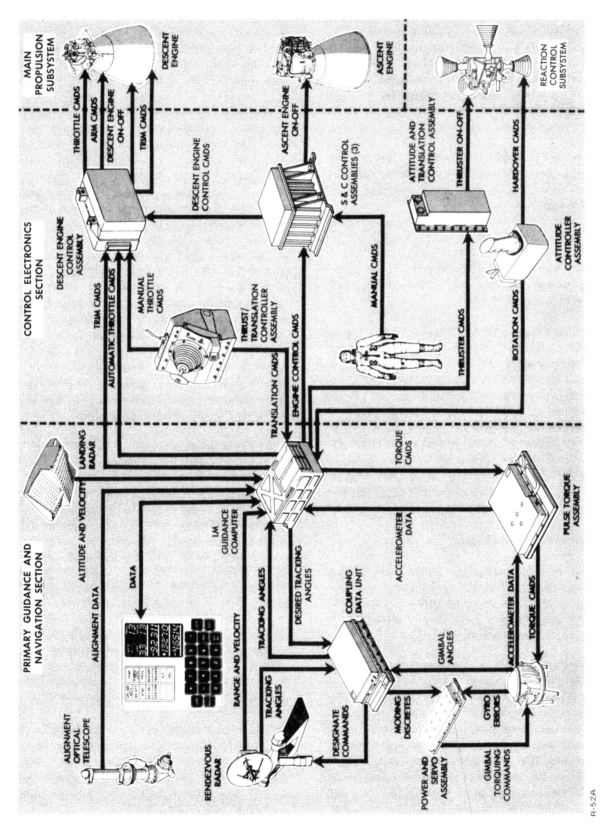

Diagram of Primary Guidance Path

R-52A

The power and servo assembly contains electronic equipment in support of the primary guidance and navigation section: power supplies for generation of internal power required by the section, servomechanisms for the inertial measurement unit, and failure detection circuitry for the inertial measurement unit.

The signal conditioner assembly provides an interface between the primary guidance and navigation section, and the Instrumentation Subsystem (IS).

OPTICAL SUBSECTION

The optical subsection consists of the alignment optical telescope and a computer control and reticle dimmer assembly.

The alignment optical telescope, an L-shaped periscope approximately 36 inches long, is used by the astronaut to take angular measurements of celestial objects. These angular measurements are required for orienting the stable member during certain periods while the LM is in flight and during prelaunch preparations while on the lunar surface. Sightings taken with the telescope are transferred to the computer by the astronaut using the computer control and reticle dimmer assembly. This assembly also controls the brightness of the telescope reticle pattern.

COMPUTER SUBSECTION

The computer subsection consists of the LM guidance computer (LGC) and a display and keyboard, which is a computer control panel. The display and keyboard is commonly referred to as "the DSKY" (pronounced "disky").

The guidance computer processes data and issues discrete control signals for various subsystems. It is a control computer with many of the features of a general-purpose computer. As a control computer, it aligns the inertial measurement unit stable member and provides rendezvous radar antenna drive commands. The LGC also provides control commands to the landing and rendezvous radars, the ascent and descent engines, the RCS thrusters, and the cabin displays. As a general purpose computer,

it solves guidance problems required for the mission. In addition, the guidance computer monitors the operation of the primary guidance and navigation section.

The guidance computer stores data pertinent to the ascent and descent flight profiles that the LM must assume to complete its mission. These data (position, velocity, and trajectory information) are used by the computer to solve flight equations. The results of various equations are used to determine the required magnitude and direction of thrust. The computer establishes corrections to be made. The LM engines are turned on at the correct time, and steering commands are controlled by the computer to orient the LM to a new trajectory, if required. The inertial subsection senses acceleration and supplies velocity changes to the computer for calculating total velocity. Drive signals are supplied from the computer to the coupling data unit and stabilization gyros in the inertial subsection to align the gimbal angles in the inertial measurement unit. Stable-member position signals are supplied to the computer to indicate attitude changes.

The computer provides drive signals to the rendezvous radar for antenna positioning and receives, from the rendezvous radar channels of the coupling data unit, antenna angle information. The computer uses this information in the antenna-positioning calculations. During lunar-landing operations, star-sighting information is manually loaded into the computer, using the DSKY. This information is used to calculate alignment commands for the inertial measurement unit. The LM guidance computer and its programming help meet the functional requirements of the mission. The functions performed in the various mission phases include both automatic and semiautomatic operations that are implemented mostly through the execution of the programs stored in the computer memory.

The DSKY provides a two-way communications link between the astronauts and the LM guidance computer. The astronauts are able to insert various parameters into the computer, display data from the computer, and to monitor data in the computer's memory.

ABORT GUIDANCE SECTION

The abort guidance section consists of an abort sensor assembly, a data entry and display assembly (DEDA), and an abort electronics assembly. The data entry and display assembly is commonly referred to as "the DEDA" (pronounced "deeda").

The abort sensor assembly, by means of gyros and accelerometers, provides incremental attitude information around the LM X, Y, and Z axes and incremental velocity changes along the LM X, Y, and Z axes. Data pulses are routed to the abort electronic assembly, which uses the LM attitude and velocity data for computation of steering errors.

The DEDA is used by the astronauts to select the desired mode of operation, insert the desired targeting parameters, and monitor related data throughout the mission. To select a mode of operation or insert data, three digits (word address) then a plus (+) or minus (–), and finally, a five digit code must be entered. If this sequence is not followed, an operator error light goes on when the enter pushbutton is pressed. To read out any parameter, three digits (address of the desired word) must be entered and a readout pushbutton pressed.

The abort electronics assembly, by means of special input-output subassemblies, interfaces the abort guidance secton with the other LM subsystems and displays. This assembly is basically a general-purpose digital computer, which solves guidance and navigation problems. Mode and submode entries coupled from the data entry and display assembly determine the operation of the computer. The computer uses incremental velocity and attitude inputs from the abort sensor assembly to calculate LM position, attitude, and velocity in the inertial reference frame. It routes altitude and altitude-rate data to altitude and altitude rate indicators; out of plane velocity data, to X-pointer indicators. Also, roll, pitch and yaw steering error signals are routed to flight director altitude indicators.

Engine-on commands are routed to the appropriate engine via the control electronics section when the following occur: an abort or abort stage pushbutton is pressed, appropriate switches are set, necessary data are entered into the DEDA, and velocity-to-be-gained exceeds a predetermined threshold (currently 2.1 fps). At the appropriate time, as determined by velocity-to-be-gained, an engine-off command is sent.

CONTROL ELECTRONICS SECTION

The control electronics section comprises two attitude controller assemblies, two thrust/translation controller assemblies, an attitude and translation control assembly, a rate gyro assembly, descent engine control assembly, three stabilization and control (S&C) control assemblies and two gimbal drive actuators.

The attitude controller assemblies are right-hand pistol grip controllers, which the astronauts use to command changes in LM attitude. These controllers function in a manner similar to an aircraft's "control stick". Each is installed with its longitudinal axis approximately parallel to LM X-axis; vehicle rotations correspond to astronaut hand movements.

The thrust/translation controller assemblies are left-hand controllers used by the astronauts to control LM translation in any axis. Vehicle translations correspond approximately to the astronauts hand movements.

The attitude and translation control assembly routes the RCS thruster on and off commands from the guidance computer to the thrusters, in the primary control mode. During abort guidance control, the assembly acts as a computer in determining which RCS thrusters are to be fired.

The rate gyro assembly is used during abort guidance control to supply the attitude and translation control assembly with damping signals to limit vehicle rotation rates and to facilitate manual rate control.

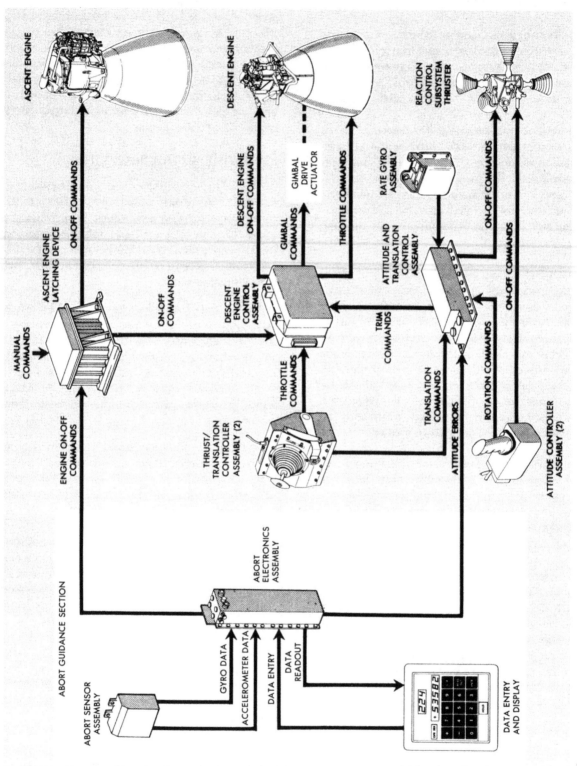

Diagram of Abort Guidance Path

R-53

GRUMMAN

The descent engine control assembly processes engine throttling commands from the astronauts (manual control) and the guidance computer (automatic control), gimbal commands for thrust vector control, preignition (arming) commands, and on and off commands to control descent engine ignition and shutdown.

The S&C control assemblies are three similar assemblies. They process, switch, and/or distribute the various signals associated with the GN&CS.

The gimbal drive actuators position the descent engine in roll and pitch in response to DECA outputs.

LANDING RADAR

The landing radar senses the velocity and slant range of the LM relative to the lunar surface by means of a three-beam Doppler velocity sensor and a single-beam radar altimeter. Velocity and range data are made available to the LM guidance computer as 15-bit binary words; forward and lateral velocity data, to the LM displays as d-c analog voltages; and range and range rate data, to the LM displays as pulse-repetition frequencies.

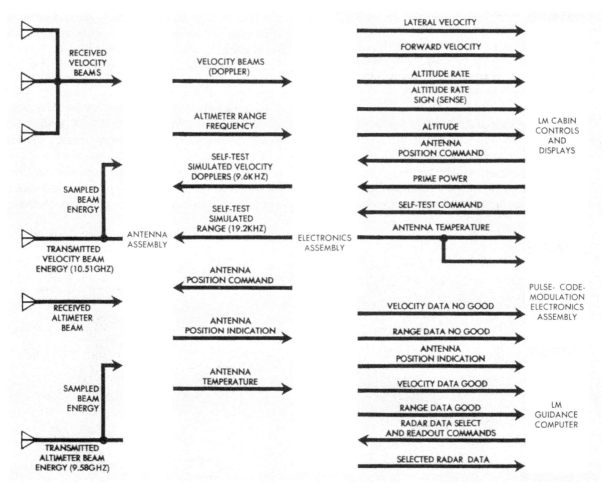

R-54

Landing Radar Signal Flow

The landing radar consists of an antenna assembly and an electronics assembly. The antenna assembly forms, directs, transmits, and receives the four microwave beams. Two interlaced phased arrays transmit the velocity- and altimeter-beam energy. Four broadside arrays receive the reflected energy of the three velocity beams and the altimeter beam. The electronics assembly processes the Doppler and continuous-wave FM returns, which provide the velocity and slant range data for the LM guidance computer and the LM displays.

The antenna assembly transmits velocity beams (10.51 gHz) and an altimeter-beam (9.58 gHz) to the lunar surface.

When the electronics assembly is receiving and processing the returned microwave beams, data-good signals are sent to the LGC. When the electronics assembly is not operating properly, data-no-good signals are sent to the pulse code modulation timing electronics assembly of the Instrumentation Subsystem for telemetry.

Using LM controls and indicators, the astronauts can monitor LM velocity, altitude, and radar-transmitter power and temperatures; apply power to energize the radar; initiate radar self-test; and place the antenna in descent or hover position. Self-test permits operational checks of the radar without radar returns from external sources. An antenna temperature control circuit, energized at earth launch, protects antenna components against the low temperatures of space environment while the radar is not operating.

The radar is first turned on and self-tested during LM checkout before separation from the CSM. The self-test circuits apply simulated Doppler signals to radar velocity sensors, and simulated lunar range signals to an altimeter sensor. The radar is self-tested again immediately before LM powered descent, approximately 70,000 feet above the lunar surface. The radar operates from approximately 50,000 feet until lunar touchdown.

Altitude (derived from slant range) is available to the LGC and is displayed on a cabin indicator at or above 25,000 feet. Slant range data are continuously updated to provide true altitude above the lunar surface. At, or above 18,000 feet, forward and lateral velocities are available to the LM guidance computer and cabin indicators.

At approximately 200 feet above the lunar surface, the LM pitches to orient its X-axis perpendicular to the surface; all velocity vectors are near zero. Final visual selection of the landing site is followed by touchdown under automatic or manual control. During this phase, the astronauts monitor altitude and velocity data from the radar.

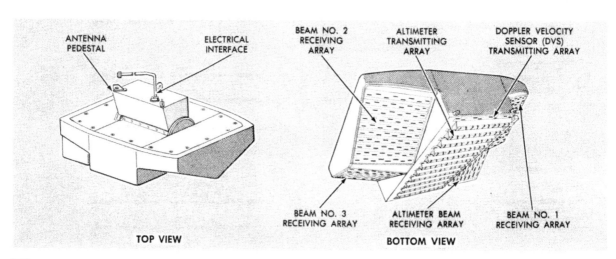

R-55

Landing Radar Antenna Assembly

The landing radar antenna has a descent position and a hover position. In the descent position, the antenna boresight angle is 24° from the LM X-axis. In the hover position, the antenna boresight is parallel to the X-axis and perpendicular to the Z-axis. Antenna position is selected by the astronaut during manual operation and by the LM guidance computer during automatic operation. During automatic operation, the LM guidance computer commands the antenna to the hover position 8,000 to 9,000 feet above the lunar surface.

RENDEZVOUS RADAR

The rendezvous radar has two assemblies, the antenna assembly and the electronics assembly. The antenna assembly automatically tracks the transponder signal after the electronics assembly acquires the transponder carrier frequency. The return signal from the transponder is received by a four-port feedhorn. The feedhorn, arranged in a simultaneous lobing configuration, is located at the focus of a Cassegrainian antenna. If the transponder is directly in line with the antenna boresight, the transponder signal energy is equally distributed to each port of the feedhorn. If the transponder is not directly in line, the signal energy is unequally distributed among the four ports.

The signal passes through a polarization diplexer to a comparator, which processes the signal to develop sum and difference signals. The sum signal represents the sum of energy received by all feedhorn ports (A + B + C + D). The difference signals, representing the difference in energy received by the feedhorn ports, are processed along two channels: a shaft-difference channel and a trunnion-difference channel. The shaft-difference signal represents the vectoral sum of the energy received by adjacent ports (A + D) – (B + C) of the feedhorn. The trunnion-difference signal represents the vectoral sum of the energy received by adjacent ports (A + B) – (C + D). The comparator outputs are heterodyned with the transmitter frequency to obtain three intermediate-frequency signals. After further processing, these signals provide unambiguous range, range rate, and direction of the CSM. This information is fed to the LGC and to cabin displays.

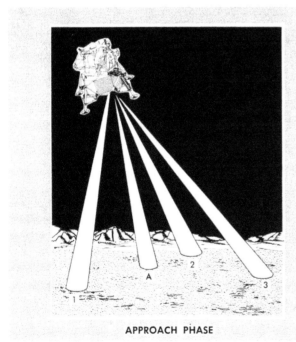

APPROACH PHASE   LANDING PHASE

R-56

Landing Radar – Antenna Beam Configuration

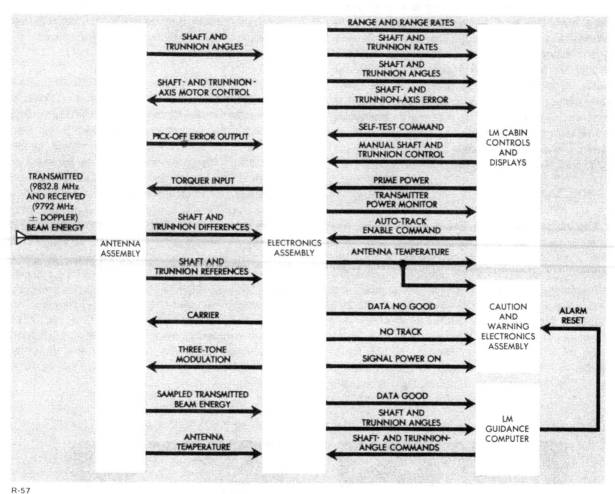

R-57

Rendezvous Radar Signal Flow

The rendezvous radar operates in three modes: automatic tracking, slew (manual), or LM guidance computer control.

Automatic Tracking Mode. This mode enables the radar to track the CSM automatically after it has been acquired; tracking is independent of LM guidance computer control. When this mode is selected, tracking is maintained by comparing the received signals from the shaft and trunnion channels with the sum channel signal. The resultant error signals drive the antenna, thus maintaining track.

Slew Mode. This mode enables an astronaut to position the antenna manually to acquire the CSM.

LM Guidance Computer Control Mode. In this mode, the computer automatically controls antenna positioning, initiates automatic tracking once the CSM is acquired, and controls change in antenna orientation. The primary guidance and navigation section, which transmits computer-derived commands to position the radar antenna, provides automatic control of radar search and acquisition.

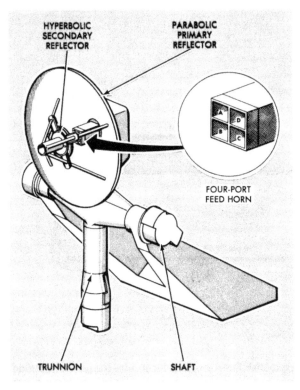

R-58

Rendezvous Radar Antenna Assembly

PRIMARY GUIDANCE PATH

The primary guidance path comprises the primary guidance and navigation section, control electronics section, landing radar, and rendezvous radar and the selected propulsion section required to perform the desired maneuvers. The control electronics section routes flight control commands from the primary guidance and navigation section and applies them to the descent or ascent engine, and the appropriate thrusters.

INERTIAL ALIGNMENT

Inertial subsection operation can be initiated automatically by the primary guidance computer or manually by the astronaut, using DSKY entries to command the computer. The inertial subsection status or mode of operation is displayed on the DSKY as determined by a computer program. When the inertial subsection is powered up, the gimbals of the inertial measurement unit are driven to zero by a reference voltage and the

coupling data unit is initialized to accept inertial subsection data. During this period, there is a 90-second delay before power is applied to the gyro and accelerometer torquing loops. This is to prevent them from torquing before the gyros reach synchronous rotor speed.

The stable member of the inertial measurement unit must be aligned with respect to the reference coordinate frame each time the inertial subsection is powered up. During flight the stable member may be periodically realigned because it may deviate from its alignment, due to gyro drift. Also, the crew may desire a new stable member orientation. The alignment orientation may be that of the CSM or that defined by the thrusting programs within the computer.

Inertial subsection alignment is accomplished in two steps: coarse alignment and fine alignment. To initiate coarse alignment, the astronaut selects, by a DSKY entry, a program that determines stable member orientation, and a coarse-alignment routine. The computer sends digital pulses, representing the required amount of change in gimbal angle, to the coupling data unit. The coupling data unit converts these digital pulses to analog signals which drive torque motors in the inertial measurement unit. As the gimbal angle changes, a gimbal resolver signal is applied to the coupling data unit, where it is converted to digital pulses. These digital pulses cancel the computer pulses stored in the coupling data unit. When this is accomplished, coarse alignment is completed and the astronaut can now select an in-flight fine-alignment routine.

To perform the fine-alignment routine, the astronaut must use the alignment optical telescope to sight on at least two stars. The gimbals, having been coarse aligned, are relatively close to their preferred angles. The computer issues fine-alignment torquing signals to the inertial measurement unit after it processes star-sighting data that have been combined with known gimbal angles.

Once the inertial subsection is energized and aligned, LM rotation is about the gimbaled stable member, which remains fixed in space. Resolvers mounted on the gimbal axes act as angle-sensing devices and measure attitude with respect to the

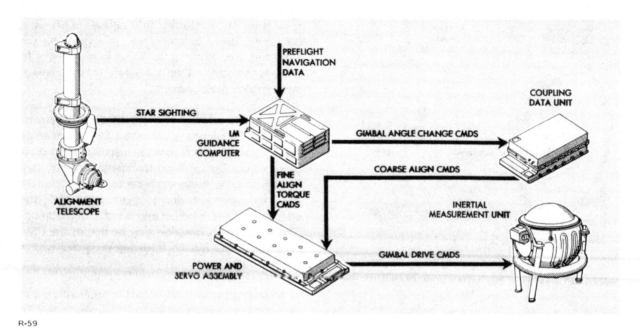

R-59

Functional Diagram of Inertial Alignment

stable member. These angular measurements are displayed to the astronauts by the flight director attitude indicators, and angular changes of the inertial reference are sent to the computer.

Inertial stability of the stable member in the inertial measurement unit is maintained with a stabilization loop which uses the IMU gyro outputs as inputs to amplifiers in the power and servo assy. The amplifier outputs drive torquers on each of the three IMU gimbals to null out the gyro errors.

ATTITUDE CONTROL

Desired attitude is calculated in the primary guidance computer and compared with the actual gimbal angles. If there is a difference between the actual and calculated angles, the inertial subsection channels of the coupling data unit generate attitude error signals, which are sent to the attitude indicators for display. These error signals are used by the digital autopilot program in the primary guidance computer to activate RCS thrusters for LM attitude correction. LM acceleration due to thrusting is sensed by three accelerometers, which are mounted on the stable member with their input axes orthogonal. The resultant signals (velocity changes) from the accelerometer loops are supplied to the computer, which calculates the total LM velocity.

Two normal modes of operation achieve attitude control: automatic and attitude hold. In addition to these two modes, there is a minimum impulse mode and a four-jet manual override mode. Either of the two normal modes may be selected on the primary guidance mode control switch.

In automatic mode, all navigation, guidance, and flight control is handled by the primary guidance computer. The computer calculates the desired or preferred attitude, generates the required thruster commands and routes them to the attitude and translation control assembly which fires the selected thruster.

Attitude hold mode is a semiautomatic mode in which either astronaut can command attitude change at an angular rate proportional to the displacement of his attitude controller. The LM holds the new attitude when the controller is brought back to its neutral (detent) position. During primary guidance control, rate commands proportional to controller displacement are sent to

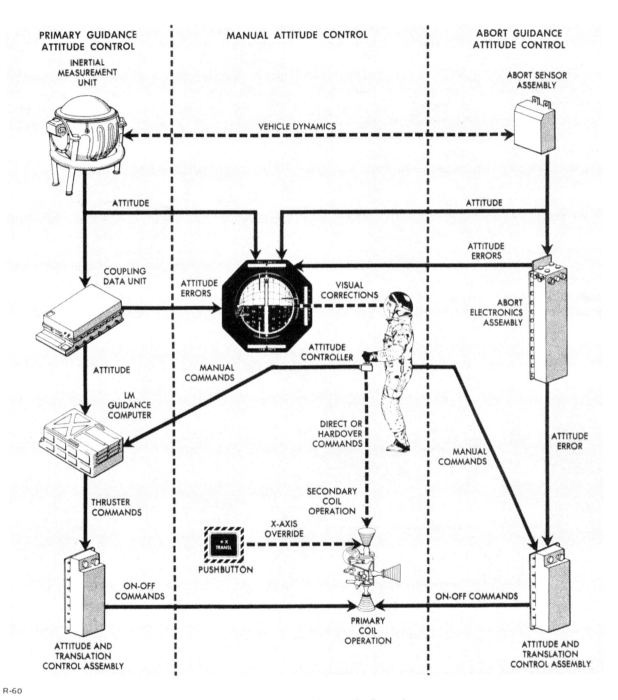

PRIMARY GUIDANCE
ATTITUDE CONTROL

MANUAL ATTITUDE CONTROL

ABORT GUIDANCE
ATTITUDE CONTROL

INERTIAL
MEASUREMENT
UNIT

ABORT SENSOR
ASSEMBLY

VEHICLE DYNAMICS

ATTITUDE

ATTITUDE

ATTITUDE
ERRORS

COUPLING
DATA UNIT

ATTITUDE
ERRORS

VISUAL
CORRECTIONS

ABORT
ELECTRONICS
ASSEMBLY

ATTITUDE
CONTROLLER

ATTITUDE

MANUAL
COMMANDS

LM
GUIDANCE
COMPUTER

DIRECT OR
HARDOVER
COMMANDS

MANUAL
COMMANDS

ATTITUDE
ERROR

THRUSTER
COMMANDS

SECONDARY
COIL
OPERATION

X-AXIS
OVERRIDE

+ X
TRANSL

PUSHBUTTON

ON-OFF
COMMANDS

ON-OFF COMMANDS

ATTITUDE AND
TRANSLATION
CONTROL ASSEMBLY

PRIMARY
COIL
OPERATION

ATTITUDE AND
TRANSLATION
CONTROL ASSEMBLY

R-60

Functional Diagram of Attitude Control

the computer. The computer processes these commands and generates thruster commands for the attitude and translation control assembly.

Minimum impulse mode enables the astronaut to control the LM with a minimum of fuel consumption. Each movement of the attitude controller out of its detent position causes the primary guidance computer to issue commands to the appropriate thrusters. The controller must be returned to the neutral position between each impulse command. This mode is selected by DSKY entry only while the control electronics section is in attitude hold. In this mode, the astronaut must perform his own rate damping and attitude steering.

Manual override also is known as the hardover mode. In certain contingencies that may require an abrupt attitude maneuver, the attitude controller

can be displaced to the maximum limit (hardover position) to command an immediate attitude change about any axis. This displacement applies signals directly to the RCS solenoids to fire four thrusters that provide the desired maneuver. This maneuver can override any other attitude control mode.

TRANSLATION CONTROL

Automatic and manual translation control is available in all three axes, using the RCS. Automatic control consists of thruster commands from the primary guidance computer to the attitude and translation control assembly. These commands are used for translations of small velocity increments and for ullage maneuvers (to settle propellant in the tanks) before ascent or descent

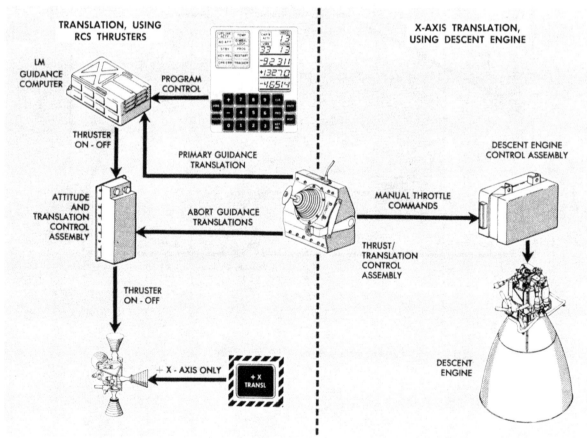

R-61

Functional Diagram of Translation Control

engine ignition after coasting phases. Manual control during primary guidance control consists of on and off commands generated by the astronaut using his thrust/translation controller. These commands are routed through the computer to the attitude and translation control assembly to fire the proper thrusters. Translation along the +X-axis can also be initiated by the astronaut using a pushbutton switch that actuates the secondary solenoid coils of the four downward firing thrusters.

DESCENT ENGINE CONTROL

Descent engine ignition is controlled either automatically by the primary guidance and navigation section, or manually through the control electronics section. Before ignition can occur, the engine arm switch must be set to the descent engine position. This opens the pre-valves to allow fuel and oxidizer to reach the propellant shutoff valves, arming the descent engine.

Engine-on commands from either computer are routed to the descent engine control assembly which commands the descent engine on by opening the propellant shutoff valves. The engine remains on until an engine-off discrete is initiated by the astronauts with either of two engine stop pushbuttons or by the computer. When the LM reaches the hover point where the lunar contact probes touch the lunar surface, a blue lunar contact light is illuminated. This indicates to the astronauts that the engine should be shut down. From this point (approximately 5 feet above the lunar surface), the LM free-falls to the lunar surface.

Descent engine throttling can be controlled by the primary guidance and navigation section and/or the astronauts. Automatic increase or decrease signals from the guidance computer are sent to the descent engine control assembly. An analog output from the control assembly corresponds to the percentage of thrust desired. The engine is controllable from 10% of thrust to a maximum of 92.5%. There are two thrust control modes: automatic and manual. In the automatic mode, the astronaut can use the selected thrust/translation controller to increase descent engine thrust only. During this mode, manual commands by the astronaut are used to override the throttle commands generated by the computer. In the manual mode, the astronauts have complete control over descent engine thrust.

Descent engine trim is automatically controlled during primary control, to compensate for center-of-gravity offsets due to propellant depletion and, in some cases for attitude control. The primary guidance computer routes trim commands for the pitch and roll axes. These signals drive a pair of gimbal drive actuators. These actuators, which are screwjack devices, tilt the descent engine about the Y-axis and Z-axis a maximum of +6° or -6° from the X-axis.

ASCENT ENGINE CONTROL

Ascent engine ignition and shutdown can be initiated automatically by the primary guidance computer or manually by the astronauts. Automatic and manual commands are routed to the S&C control assemblies. These assemblies provide logically ordered control of LM staging and engine on and off comands. The control assemblies are enabled when the astronauts select the ascent engine position of the engine arm switch.

In an abort stage situation while the descent engine is firing, the control assemblies provide a time delay before commanding staging and ascent engine ignition. The time delay ensures that descent engine thrusting has completely stopped before staging occurs.

ABORT GUIDANCE PATH

The abort guidance path comprises the abort guidance section, control electronics section, and the selected propulsion section. The abort guidance path performs all inertial guidance and navigation functions necessary to effect a safe orbit or rendezvous with the CSM. The stabilization and control functions are performed by analog computation techniques, in the control electronics section.

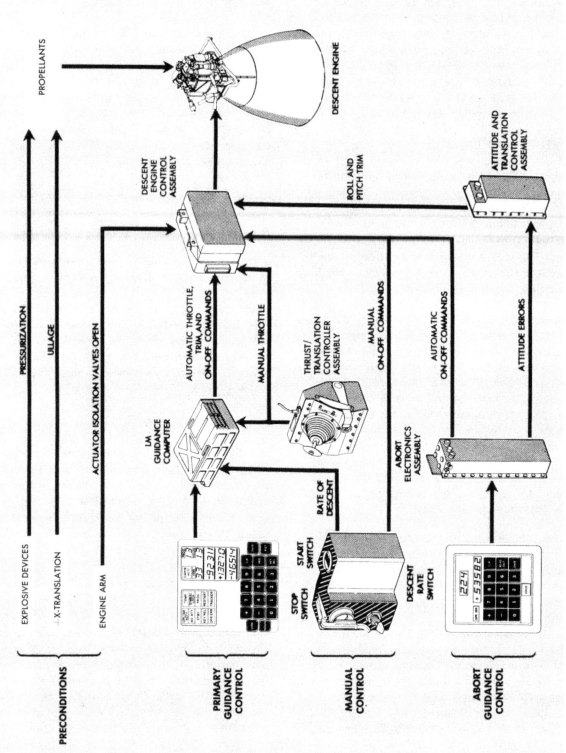

Functional Diagram of Descent Engine Control

R-62

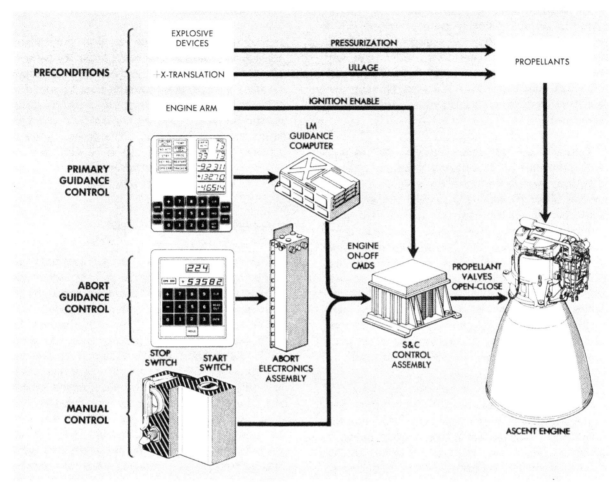

R-63

Functional Diagram of Ascent Engine Control

The control electronics section functions as an autopilot when the abort guidance path is selected. It uses inputs from the abort guidance section and from the astronauts to provide the following: on, off, and manual throttling commands for the descent engine; descent engine gimbal drive actuator commands; ascent engine on and off commands; engine sequencer logic to ensure proper arming and staging before engine startup and shutdown; RCS on and off commands; RCS jet-select logic to select the proper thruster for the various maneuvers; and modes of control, ranging from automatic to manual.

ATTITUDE CONTROL

The abort guidance path operates in the automatic mode or the attitude hold mode. In automatic, navigation and guidance functions are controlled by the abort guidance section, attitude by the control electronics section. The abort electronics assembly (abort guidance computer) generates roll, pitch, and yaw attitude error signals, which are summed with rate-damping and attitude rate signals in the attitude and translation control assembly. A jet-select logic circuit selects the thruster to be fired and issues the appropriate thruster command.

In attitude hold, the astronaut uses manual control. In this mode, a pulse submode and a two-jet direct submode are available in addition to manual override (hardover). The pulse and two-jet direct submodes are selectable on an individual axis basis only. The attitude controller generates attitude rate, pulse, direct, and hardover commands.

During abort guidance control, with the attitude controller in the neutral position, attitude is held by means of attitude error signals detected by the abort electronics assembly. When either controller is moved out of the neutral position, the attitude error signals from the abort guidance section are zero. Rate commands, proportional to controller displacement, are processed in the attitude and translation control assembly, and the thrusters are fired until the desired vehicle rate is achieved. When the controller is returned to the neutral position, the vehicle rate is reduced to zero and the abort guidance section holds the LM in the new attitude.

The pulse submode is selected by the astronaut, using the appropriate attitude control switch. Automatic attitude control about the selected axis is then disabled and a fixed train of pulses is generated when the attitude controller is displaced from its neutral position. To change vehicle attitude in this submode, the attitude controller must be moved out of neutral. This commands acceleration about the selected axis through low-frequency thruster pulsing. The pulse submode uses the primary solenoid coils of the thrusters; the direct submode, the secondary solenoid coils. To terminate rotation, an opposite acceleration about the selected axis must be commanded.

The direct submode is selected by the astronaut, using the attitude control switches that are used for the pulse submode. When selected, automatic control about the selected axes is disabled and direct commands are routed to the RCS secondary solenoids to two thrusters when the attitude controller is displaced from the neutral position. The thrusters under direct control fire continuously until the controller is returned to the neutral position.

TRANSLATION CONTROL

During abort guidance control, only manual translation is available because the abort programs do not require lateral or forward translation maneuvers. Translation control consists of on and off commands from a thrust/translation controller to the jet select logic of the attitude and translation control assembly. RCS thrust along the +X-axis is accomplished the same way as during primary guidance control when the astronaut uses the +X-axis translation pushbutton.

DESCENT ENGINE CONTROL

Descent engine ignition is automatically controlled by programs stored in the abort electronics assembly. This assembly computes the abort guidance trajectory and required steering. If the primary guidance and navigation section fails while the descent engine is being used, the astronaut initiates abort guidance descent engine control through a DEDA entry. The abort electronics assembly can only control descent engine ignition and shutdown. Descent engine throttling and gimbaling are not under computer control when operating with the abort guidance section. As with the primary guidance path, the abort path generates an engine-off command when the required velocity is attained. This velocity depends upon whether the program used will place the LM in a rendezvous trajectory or in a parking orbit. Manual on and off control also is available. In all cases, the S&C control assemblies receive engine on and off commands. As in the primary guidance path, these assemblies route the commands to the descent engine control assembly which routes them to the engine.

The astronaut uses the thrust/translation controller to control descent engine throttling and translation maneuvers. The manual throttle commands are supplied to the descent engine control assembly, which generates analog signals driving the throttle valve actuator.

Descent engine trim control under abort guidance, is achieved by using attitude errors from the abort electronics assembly. These errors are used by the attitude and translation control assembly for attitude control and steering calculation. The roll and pitch attitude errors are routed to the descent engine control assembly as trim commands.

ASCENT ENGINE CONTROL

Ascent engine control during abort guidance is similar to that of the primary guidance. During abort guidance control, automatic ascent engine ignition and shutdown are controlled by the abort electronics assembly.

If the descent stage is attached, the LM can be staged manually through use of the appropriate switches on the explosive devices panel. The astronaut has the option of using an abort stage pushbutton to start an automatic ascent engine ignition sequence. If the ascent engine-on command is lost, the ascent engine latching device memory circuit keeps issuing the command.

EQUIPMENT

PRIMARY GUIDANCE AND NAVIGATION SECTION

NAVIGATION BASE

The navigation base is a lightweight mount (about 3 pounds) bolted to the LM structure above the astronauts heads, with three mounting pads on a center ring. The center ring is approximately 14 inches in diameter and each of the four legs, which are part of the base, is approximately 10 inches long.

INERTIAL MEASUREMENT UNIT

The inertial measurement unit contains the stable member, gyroscopes, and accelerometers necessary to establish the inertial reference.

R-64

Abort Stage Functions

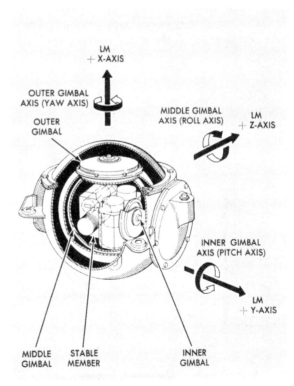

R-65

Inertial Measurement Unit

The stable member serves as the space-fixed reference for the inertial subsection. It is supported by three gimbal rings (outer, middle, and inner) for complete freedom of motion.

The outer gimbal is mounted to the case of the unit; its axis is parallel to the LM X-axis. The middle gimbal is mounted to and perpendicular with the outer; its axis is parallel to the LM Z-axis. The inner gimbal supports the stable member; its axis is parallel to the LM Y-axis. The inner gimbal is mounted to the middle one. All three gimbals are spherical with 360 degrees of freedom. To overcome the small amount of friction inherent in the support system, small torque motors are mounted on each axis.

The three Apollo inertial reference integrating gyroscopes, used to sense attitude changes, are mounted on the stable member, mutually perpendicular. The gyros are fluid- and magnetically-suspended, single-degree-of-freedom types. They sense displacement of the stable member and generate error signals proportional to displacement. The three pulse integrating pendulous accelerometers are fluid- and magnetically-suspended devices.

Thermostats maintain gyro and accelerometer temperature within their required limits during inertial measurement unit standby and operating modes. Heat is applied to end-mount heaters on the inertial components, by stable member heaters, and by a temperature control anticipatory heater. Heat is removed by convection, conduction, and radiation. The natural convection used during inertial measurement unit standby mode is changed to blower-controlled, forced convection during the operating mode. Inertial measurement unit internal pressure is normally between 3.5 and 15 psia, enabling the required forced convection. To aid in removing heat, water-glycol passes through the case. Therefore, heat flow is from the stable member to the case and coolant. The temperature control system consists of the temperature control circuit, the blower control circuits, and temperature alarm circuit.

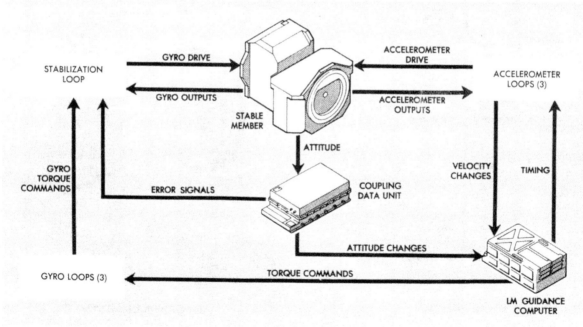

R-66

Inertial Subsection Functional Loops

COUPLING DATA UNIT

The coupling data unit performs analog-to-digital conversion, digital-to-analog conversion, inertial subsection moding and failure detection. It consists of a sealed container which encloses 34 modules of 10 different types that make up five almost identical channels: one each for the inner, middle, and outer gimbals of the inertial measurement unit and one each for the rendezvous radar shaft and trunnion gimbals. Several of the modules are shared by all five channels.

The two channels used with the rendezvous radar interface between the antenna and the guidance computer. The computer calculates digital antenna position commands before acquisition of the CSM. These signals are converted to analog form by the coupling data unit and applied to the antenna drive mechanism to aim the antenna. Tracking-angle information in analog form is converted to digital by the unit and applied to the guidance computer.

The three channels used with the inertial measurement unit provide interfaces between it and the guidance computer and between the computer and the abort guidance section. Each of the three IMU gimbal angle resolvers provide its channel with analog gimbal-angle signals that represent LM attitude. The coupling data unit converts these signals to digital form and applies them to the guidance computer. The computer calculates attitude or translation commands and routes them through the control electronics section to the proper thruster. The coupling data unit converts attitude error signals to 800-cps analog signals and applies them to the attitude indicator. Coarse- and fine-alignment commands generated by the guidance computer are coupled to the inertial measurement unit through the coupling data unit.

The digital-to-analog converters of the coupling data unit are a-c ladder networks. When the unit is used to position a gimbal, the guidance computer calculates the difference between the desired gimbal angle and the actual gimbal angle. This difference results in a servo error signal that drives the gimbal to the desired angle.

The analog-to-digital converter operates on an incremental basis. Using a digital-analog feedback technique which utilizes the resolvers as a reference, the coupling data unit accumulates the proper angular value by accepting increments of the angle to close the feedback loop. These data are applied to counters in the guidance computer for rendezvous radar tracking information, and to counters in the primary and abort guidance computers for the inertial reference gimbal angles. In this manner, the abort guidance section attitude reference is fine-aligned simultaneously with that of the primary guidance and navigation section.

PULSE TORQUE ASSEMBLY

The pulse torque assembly consists of 17 electronic modular subassemblies mounted on a common base. There are four binary current switches: one furnishes torquing current to the three gyros; the other three furnish torquing current to the three accelerometers. Four d-c differential amplifier and precision voltage reference subassemblies regulate torquing current supplied through the binary current switches.

Three a-c differential amplifier and interrogator subassemblies amplify accelerometer signal generator signals and convert them to positive and negative torque pulses. The gyro calibration module applies torquing current to the gyros when commanded by the guidance computer. Three accelerometer calibration modules compensate for the difference in inductive loading of accelerometer torque generator windings and regulate the balance of positive and negative torque. A pulse torque isolation transformer couples torque commands, data pulses, interrogate pulses, switching pulses, and synchronizing pulses between the guidance computer and the pulse torque assembly. The pulse torque power supply supplies power for the other 16 subassemblies.

POWER AND SERVO ASSEMBLY

The power and servo assembly provides a central mounting point for the primary guidance and navigation section amplifiers, modular electronic components, and power supplies. The assembly is on the cabin bulkhead behind the astronauts. It consists of 14 subassemblies mounted to a header assembly.

SIGNAL CONDITIONER ASSEMBLY

The signal conditioner assembly preconditions primary guidance and navigation section measurements to a 0- to 5-volt d-c format before the signals are routed to the Instrumentation Subsystem.

ALIGNMENT OPTICAL TELESCOPE

The alignment optical telescope, mounted on the navigation base to provide mechanical alignment and a common reference between the telescope and the inertial measurement unit, is a unity-power, periscope-type device with a 60° conical field of view. It is operated manually by the astronauts. The telescope has a movable shaft axis (parallel to the LM X-axis) and a line of sight approximately 45° from the X-axis in the Y-Z plane.

The telescope line of sight is fixed in elevation and movable in azimuth to six detent positions. These detent positions are selected by turning a detent selector knob on the telescope; they are located at 60° intervals. The forward (F), zero detent position, places the line of sight in the X-Z plane. looking forward and up as one would look from inside the LM. The right (R) position places the line of sight 60° to the right of the X-Z plane; the left (L) position, 60° to the left of the X-Z plane. Each of these positions maintains the line of sight at 45° from the LM +X-axis. The remaining three detent positions reverse the prism on top of the telescope. These positions are right-rear, closed (CL), and left-rear. The CL position (180° from the F position) is the stowed position. The right-rear and left-rear positions have minimal use.

The optics consist of two sections: shaft optics and eyepiece optics. The shaft optics section is a -5 power complex that provides a 60° field of view. The eyepiece optics section is a +5 power complex that provides shaft and trunnion angle

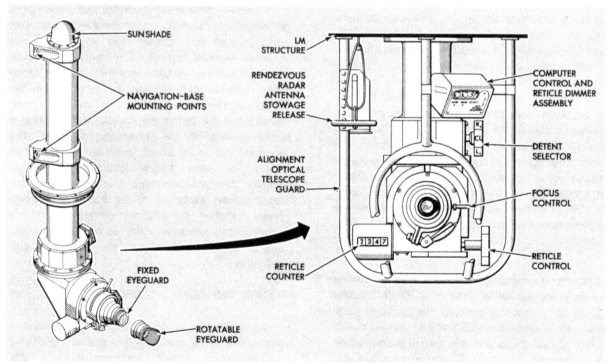

R-67

Alignment Optical Telescope

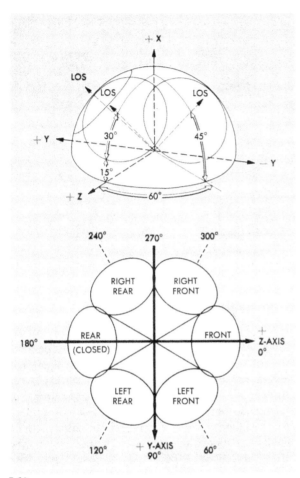

R-68

Alignment Optical Telescope – Detents and Field of View

measurements. The reticle pattern within the eyepiece optics consists of crosshairs and a pair of Archimedes spirals. The vertical crosshair, an orientation line designated the Y-line, is parallel to the LM X-axis when the reticle is at the 0° reference position. The horizontal crosshair, an auxiliary line designated the X-line, is perpendicular to the orientation line. The one-turn spirals are superimposed from the center of the field of view to the top of the vertical crosshair. Ten miniature red lamps mounted around the reticle prevent false star indications caused by imperfections in the reticle and illuminate the reticle pattern. Stars will appear white; reticle imperfections, red. Heaters prevent fogging of the mirror due to moisture and low temperatures during the mission.

A reticle control enables manual rotation of the reticle for use in lunar surface alignments. A counter on the left side of the unit, provides angular readout of the reticle rotation. The counter reads in degrees to within ±0.02° or ±72 seconds. The maximum reading is 359.88°, then the counter returns to 0°. Interpolation is possible to within ±0.01°.

A rotatable eyeguard is fastened to the end of the eyepiece section. The eyeguard is axially adjustable for head position. It is used when the astronaut takes sightings with his faceplate open. This eyeguard is removed when the astronaut takes sightings with his faceplate closed; a fixed eyeguard, permanently cemented to the telescope, is used instead. The fixed eyeguard prevents marring of the faceplate by the eyepiece. A high-density filter lens, supplied as auxiliary equipment, prevents damage to the astronaut's eyes due to accidental direct viewing of the sun or if the astronaut chooses to use the sun as a reference.

The alignment optical telescope is used for in-flight and lunar surface sightings.

For in-flight sightings, the telescope may be placed in any of the usable detent positions. However, when the LM is attached to the CSM, only the forward position is used. The astronaut selects a detent and the particular star he wishes to use. He then maneuvers the LM so that the selected star falls within the telescope field of view. The specific detent position and a code associated with the selected star are entered into the guidance computer by the astronaut using the DSKY. The LM is then maneuvered so that the star image crosses the reticle crosshairs. When the star image is coincident with the Y-line, the astronaut presses the mark Y pushbutton; when it is coincident with the X-line, he presses the mark X pushbutton. The astronaut may do this in either order and, if desired, he may erase the latest mark by pressing the reject pushbutton. When a mark pushbutton is pressed, a discrete is sent to the guidance computer. The guidance computer then records the time of mark and the inertial measurement unit gimbal angles at the instant of the mark.

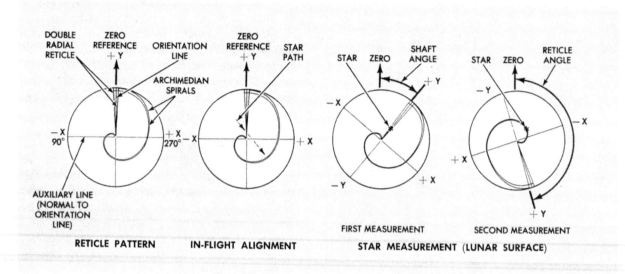

R-69

Alignment Optical Telescope – Reticle Pattern

Crossing of a reticle line by the star image defines a plane containing the star. Crossing of the other reticle line defines another plane containing the same star. The intersection of these planes forms a line that defines the direction of the star. To define the inertial orientation of the stable member, sightings on at least two stars are required. Each star sighting requires the same procedure. Multiple reticle crossings and their corresponding marks can be made on either or both stars to improve the accuracy of the sightings. Upon completion of the second star sightings, the guidance computer calculates the orientation of the stable member with respect to a predefined reference coordinate system.

On the lunar surface, the LM cannot be maneuvered to obtain a star-image that crosses the reticle crosshairs. The astronaut using the reticle control knob, adjusts the reticle to superimpose the orientation (Y) line on the target star. The reticle angle display on the reticle counter, is then inserted into the computer by the astronaut. This provides the computer with the star orientation angle (shaft angle). The astronaut then continues rotating the reticle until a point on the spirals is superimposed on the target star. This second angular readout

(reticle angle) is then entered into the computer along with the detent position and the code of the observed star. The computer can now calculate the angular displacement of the star from the center of the field of view by computing the difference between the two counter readings. Due to the characteristics of the reticle spirals, the Δ angle is proportional to the distance of the star from the center of the field of view. Using this angle and a proportionality equation, the computer can calculate the trunnion angle. At least two star sightings are required for determination of the inertial orientation of the stable member.

COMPUTER CONTROL AND RETICLE DIMMER ASSEMBLY

The computer control and reticle dimmer assembly is mounted on the alignment optical telescope guard. The mark X and mark Y pushbuttons are used by the astronauts to send discrete signals to the primary guidance computer when star sightings are made. The reject pushbutton is used if an invalid mark has been sent to the computer. A thumbwheel on the assembly is used to adjust the brightness of the telescope's reticle lamps.

LM GUIDANCE COMPUTER

The LM guidance computer is the central data-processing device of the GN&CS. It is a parallel fixed-point, one's-complement, general-purpose digital computer with a fixed rope core memory and an erasable ferrite-core memory. It has a limited self-check capability. Inputs to the computer are received from the landing radar and rendezvous radar, from the inertial measurement unit through the inertial channels of the coupling data unit and from an astronaut through the DSKY. The computer performs four major functions: (1) calculates steering signals and generates engine and RCS thruster commands to keep the LM on a required trajectory (2) aligns the stable member (inner gimbal) of the inertial measurement unit to a coordinate system defined by precise optical measurements, (3) conducts limited malfunction isolation for the GN&CS, and (4) computes pertinent navigation information for display to the astronauts. Using information from navigation fixes, the computer determines the amount of deviation from the required trajectory and calculates the necessary attitude and thrust corrective commands. Velocity corrections are measured by the inertial measurement unit and controlled by the computer. During coasting phases of the mission, velocity corrections are not made continuously, but are initiated at predetermined checkpoints.

The computer's memory consists of an erasable and a fixed magnetic core memory with a combined capacity of 38,916 16-bit words. The erasable memory is a coincident-current, ferrite core array with a total capacity of 2,048 words; it is characterized by destructive readout. The fixed memory consists of three magnetic-core rope modules. Each module contains two sections; each section contains 512 magnetic cores. The capacity of each core is 12 words, making a total of 36,864 words in the fixed memory. Readout from the fixed memory is non-destructive.

The logic operations of the computer are mechanized using micrologic elements, in which the necessary resistors are diffuzed into single silicon wafers. One complete NOR gate, which is the basic building block for all the circuitry, is in a package the size of an aspirin tablet. Flip-flops, registers, counters, etc. are made from these standard NOR elements in different wiring configurations. The computer performs all necessary arithmetic operations by addition, adding two complete words and preparing for the next operation in approximately 24 microseconds. To subtract, the computer adds the complement of the subtrahend. Multiplication is performed by successive additions and shifting; division, by successive addition of complements and shifting.

Functionally, the computer contains a timer, sequence generator, central processor, priority control, an input-output section, and a memory unit.

The timer generates all necessary synchronization pulses to ensure a logical data flow with the LM subsystems. The sequence generator directs the execution of the programs. The central processor performs all arithmetic operations and checks information to and from the computer. Memory stores the computer data and instructions. Priority control establishes a processing priority for operations that must be performed by the computer. The input output section routes and conditions signals between the computer and the other subsystems.

The main functions of the computer are implemented through execution of programs stored in memory. Programs are written in machine language called basic instructions. A basic instruction can be an instruction word or a data word. Instruction words contain a 12-bit address code and a three-bit order code.

The computer operates in an environment in which many parameters and conditions change in a continuous manner. The computer, however, operates in an incremental manner, one item at a time. Therefore, for it to process the parameters, its hardware is time shared. The time sharing is accomplished by assigning priorities to the processing functions. These priorities are used by the computer so that it processes the highest priority processing function first.

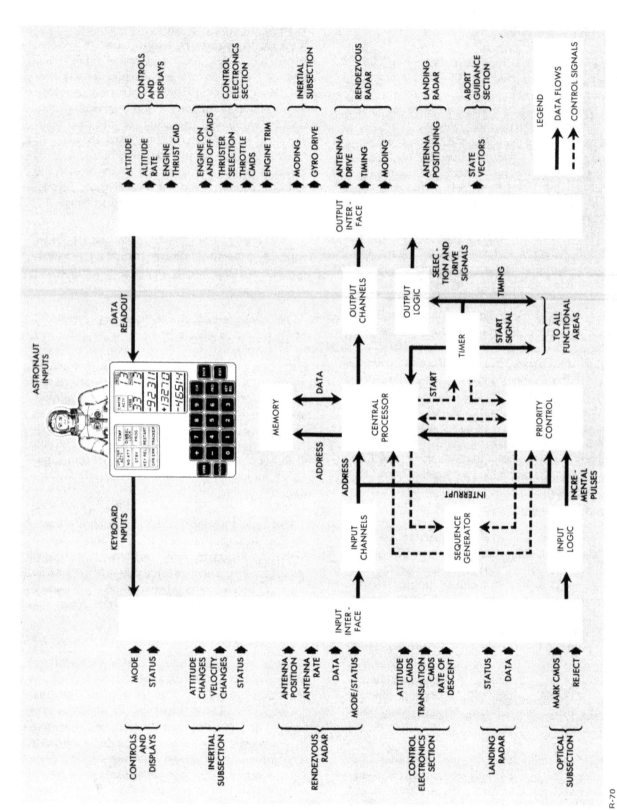

Diagram of LM Guidance Computer

In addition, each of the functions has a relative priority with respect to the others; also within each there are a number of processing functions, each having a priority level relative to the other in the group. Most of the processing performed by the computer is in the program controlled processing category. During this processing the computer is controlled by the program stored in its memory.

Real time, which is used in solving guidance and navigation problems, is maintained within the computer's memory. A 745.65-hour (approximately 31 days) clock is provided. The clock is synchronized with ground elapsed time (GET) which is "time zero" at launch. This time is transmitted once every second by downlink operation for comparison with MSFN elapsed time.

Incremental transmissions occur in the form of pulse bursts from the output channels to the coupling data unit, the gyro fine-alignment electronics, the RCS, and the radars. The number of pulses and the time at which they occur are controlled by the program. Discrete outputs, originating in the output channels under program control, are sent to the DSKY and other subsystems. A continuous pulse train at 1.024 mHz originates in the timing output logic and is sent as a synchronization signal to the timing electronics assembly in the Instrumentation Subsystem (IS).

The uplink word from MSFN via the digital uplink assembly is supplied as an incremental pulse to the priority control. As this word is received, priority produces the address of the uplink counter in memory and requests the sequence generator to execute the instructions that perform the serial-to-parallel conversion of the input word. When the conversion is completed, the parallel word is transferred to a storage location in memory by the uplink priority program. The uplink priority program also retains the parallel word for subsequent downlink transmission. Another program converts the parallel word to a coded display format and transfers the display information to the DSKY.

The downlink operation is asynchronous with respect to the IS. The IS supplies all the timing signals necessary for the downlink operation.

Through the DSKY, the astronaut can load information into the computer, retrieve and display information contained in the computer, and initiate any program stored in memory. A key code is assigned to each keyboard pushbutton. When a DSKY pushbutton is pressed, the key code is sent to an input channel of the computer. A number of key codes are required to specify an address or a data word. The initiated program also converts the keyboard information to a coded display format, which is transferred by another program to an output channel and to the DSKY for display. The display is a visual indication that the key code was received, decoded, and processed properly.

DISPLAY AND KEYBOARD

The DSKY is located on panel 4 between the Commander and LM Pilot and above the forward hatch. The upper half is the display portion; the lower half comprises the keyboard. The display portion contains seven caution indicators, seven status indicators, seven operation display indicators, and three data display indicators. These displays provide visual indications of data being loaded in the computer, the computer's condition and the program being used. The displays also provide the computer with a means of displaying or requesting data.

The caution indicators when on, are yellow; the status indicators, white. The operation and data displays are illuminated green when energized. The words "PROG," "VERB," and "NOUN" and the lines separating the three groups of display indicators, and the 19 pushbuttons of the keyboard are illuminated when the guidance computer is powered-up.

Pushbutton	Function
0 through 9	Enters numerical data, noun codes, and verb codes into computer
+ and −	Informs computer that following numerical data are decimal and indicates sign of data
VERB	Indicates to computer that it is going to take some action and conditions computer to interpret the next two numerical characters as a verb code
NOUN	Conditions computer to interpret next two numerical characters (noun code) as to what type of action is applied to verb code
CLEAR	Clears data contained in data display; pressing this pushbutton clears data display currently being used. Successive pressing clears other two data displays
PRO	Commands computer to proceed to standby mode; if in standby mode, commands computer to resume regular operation
KEY REL	Releases keyboard displays initiated by keyboard action so that information supplied by computer program may be displayed
ENTR	Informs computer that data to be inserted is complete and that requested function is to be executed
RSET	Turns off condition indicator lamps after condition has been corrected

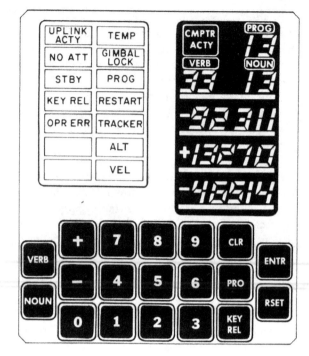

R-71A

Display and Keyboard

The DSKY enables the astronauts to insert data into the guidance computer and to initiate computer operations. The astronauts can also use the keyboard to control the moding of the inertial subsection. The exchange of data between the astronauts and the computer is usually initiated by an astronaut; however, it can also be initiated by internal computer programs.

The operator of the DSKY can communicate with the computer by pressing a sequence of pushbuttons on the DSKY keyboard. The computer can also initiate a display of information or request the operator for some action, through the processing of its program.

The basic language between the astronaut and the DSKY consists of verb and noun codes. The verb code indicates what action is to be taken (operation). The noun code indicates to what this action is applied (operand). Verb and noun codes may be originated manually or by internal computer sequence. Each verb or noun code contains two numerals. The standard procedure for manual operation involves pressing a sequence of seven pushbuttons:

$$\text{VERB} \quad V_1 \quad V_2 \quad \text{NOUN} \quad N_1 \quad N_2 \quad \text{ENTR}$$

Pressing the verb pushbutton blanks the verb code display on the display panel and clears the verb code register within the computer. The next two pushbuttons (0 to 9) pressed provide the verb code (V_1 and V_2). Each numeral of the code is displayed by the verb display as the pushbutton is pressed. The noun pushbutton operates the same as the verb pushbutton, for the noun display and noun code register. The enter pushbutton starts the operation called for. It is not necessary to follow any order in punching in the verb or noun code. It can be done in reverse order, and a previously entered verb or noun may be used without repunching it.

An error noticed in the verb code or the noun code before pressing the enter pushbutton is corrected by pressing the verb or noun pushbutton and repunching the erroneous code, without changing the other one. Only when the operator has verified that the desired verb and noun codes are displayed does he press the enter pushbutton.

Decimal data are identified by a plus or minus sign preceding the five digits. If a decimal format is used for loading data, it must be used for all components of the verb. Mixing of decimal and octal data for different components of the same load verb is not permissible. If data are mixed, the OPR ERR condition light goes on.

After any use of the DSKY, the numerals (verb, noun, and data words) remain visible until the next use of the DSKY. If a particular use of the DSKY involves fewer than three data words, the unused data display registers remain unchanged unless blanked by deliberate program action. Some verb-noun codes require additional data to be loaded. If additional data are required after the enter pushbutton is pressed, following the keying of the verb-noun codes, the verb and noun displays flash on and off at a 1.5-Hz rate. These displays continue to flash until all information associated with the verb-noun code is loaded.

OPERATION UNDER COMPUTER CONTROL

Keyboard operations by the internal computer sequences are the same as those described for manual operation. Computer-initiated verb-noun combinations are displayed as static or flashing displays. A static display identifies data displayed only for astronaut information; no crew response is required. A flashing display calls for appropriate astronaut response as dictated by the verb-noun combination. In this case, the internal sequence is interrupted until the operator responds appropriately, then the flashing stops and the internal sequence resumes. A flashing verb-noun display must receive only one of the proper responses, otherwise, the internal sequence that instructed the display may not resume.

ABORT GUIDANCE SECTION

ABORT SENSOR ASSEMBLY

This assembly contains three floated, pulse-rebalanced, single-degree-of-freedom, rate-integrating gyros and three pendulous reference accelerometers. These six sensors are aligned with the three LM reference axes and housed in a beryllium block mounted on the navigation base. The assembly is controlled to maintain its internal temperature at $+120^\circ$ F, with external temperatures between -65° and $+185^\circ$ F. This is accomplished by two temperature control circuits, one each for fast warmup and fine temperature control. During fast warmup, temperature can be raised from 0° to $+116^\circ$ F in 40 minutes. The fine temperature control circuit controls the temperature after $+116^\circ$ F is reached and raises the temperature 4°. This operating temperature ($+120^\circ$ F) is maintained within 0.20° F.

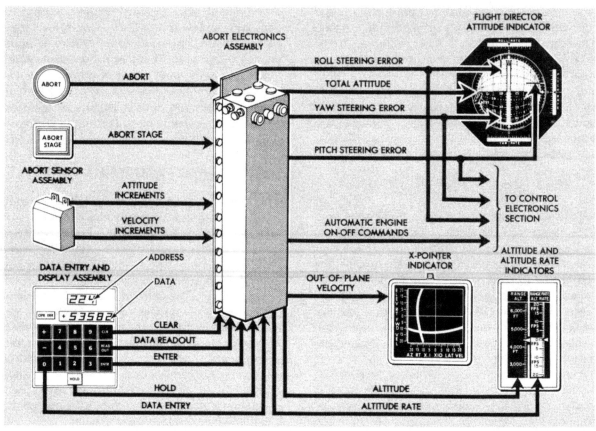

R-72

Diagram of Abort Guidance Section

DATA ENTRY AND DISPLAY ASSEMBLY

Essentially, the DEDA consists of a control panel to which electroluminescent displays and data entry pushbuttons are mounted and a logic enclosure that houses logic and input/output circuits.

As each numerical pushbutton is pressed, its code is displayed. When the appropriate number of pushbuttons are pressed, the enter or readout pushbutton can be pressed to complete the operation. The logic circuits process octal and decimal data. Octal data consists of a sign and five octal characters. Decimal data consists of a sign and five binary-coded decimal characters. The input/output circuits transfer data to and from the abort electronics assembly (computer). Data transfer occurs when the computer detects the depression of the enter or readout pushbutton.

ABORT ELECTRONICS ASSEMBLY

This assembly is a high-speed, general-purpose computer with special-purpose input/output electronics. It uses a fractional two's complement, parallel arithmetic section and parallel data transfer. Instruction words are 18 bits long; they consist of a five-bit order code, an index bit, and a 12-bit operand address. For purposes of explanation, the assembly may be separated into a memory, central computer, and input/output subassembly.

The memory is a coincident-current, parallel, random-access, ferrite-core stack with a capacity of 4,096 instruction words. It is divided into two sections: temporary storage and permanent storage. Each section has a capacity of 2,048 instruction words. The temporary memory stores replaceable instructions and data. Temporary results may be stored in this memory and may be updated as necessary. The permanent memory stores instructions and constants that are not modified during a mission. The cycle time of the memory is 5 microseconds.

Basically, the central computer consists of eight data and control registers, two timing registers, and associated logic. The data and control registers are interconnected by a parallel data bus. Central computer operations are executed by appropriately timed transfer, controlled by the timing registers, of information between the registers, memory, and input/output subassembly.

The input/output subassembly consists of four basic types of registers: integrator, ripple counter, shift, and static. These registers operate independently of the central computer, except when they are accessed during execution of an input or output instruction. All transfers of data between the central computer and the input-output registers are in parallel.

CONTROL ELECTRONICS SECTION

ATTITUDE CONTROLLER ASSEMBLIES

Each attitude controller assembly supplies attitude rate commands proportional to the displacement of its handle, to the computer and the attitude and translation control assembly; supplies an out-of-detent discrete each time the handle is out of its neutral position; and supplies a followup discrete to the abort guidance section each time the controller is out of detent. A trigger-type push-to-talk switch on the pistol grip handle of the controller assembly is used for communication with the CSM and ground facilities.

As the astronaut uses his attitude controller, his hand movements are analogous to vehicle rotations. Clockwise or counterclockwise rotation of the controller commands yaw right or yaw left, respectively. Forward or aft movement of the controller commands vehicle pitch down or up, respectively. Left or right movement of the controller commands roll left or right, respectively.

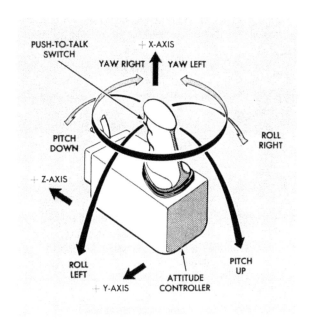

R-73

Attitude Controller Assembly Manipulations

Each assembly consists of position-sensing transducers, out-of-detent switches, and limit switches installed about each axis. The transducers provide attitude rate command signals that are proportional to controller displacements. The out-of-detent switches provide pulsed or direct firing of the thrusters when either mode is selected. The limit switches are wired to the secondary solenoid coils of the thrusters. Whenever the controller is displaced to its hardstops (hardover position), the limit switches close to provide commands that override automatic attitude control signals from the attitude and translation control assembly.

THRUST/TRANSLATION CONTROLLER ASSEMBLIES

The thrust/translation controller assemblies are functionally integrated translation and thrust controllers. The astronauts use these assemblies to command vehicle translations by firing RCS thruster and to throttle the descent engine between 10% and 92.5% thrust magnitude. The controllers are three axis, T-handle, left-hand controllers; they are mounted with their longitudinal axis approximately 45° from a line parallel to the LM Z-axis (forward axis).

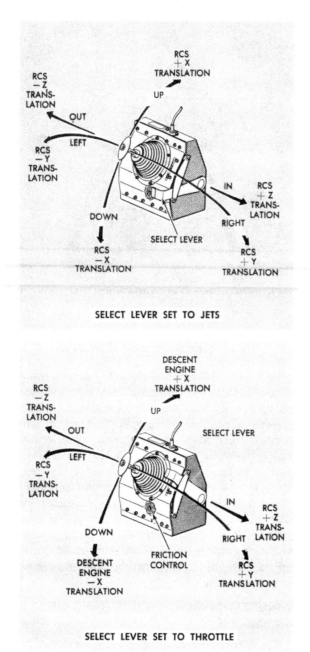

SELECT LEVER SET TO JETS

SELECT LEVER SET TO THROTTLE

R-74

Thrust / Translation Controller Assembly Manipulations

Setting a switch in the LM cabin determines whether the Commander's or LM Pilot's assembly is in command. A lever on the right side of the controller enables the astronaut to select either of two control functions: (1) to control translation in the Y-axis and Z-axis using the RCS thrusters and throttling of the descent engine to control X-axis translation; and (2) to control translation in all three axes using the RCS thrusters.

Due to the assembly mounting position, LM translations correspond to astronaut hand movements when operating the controller. Moving the T-handle to the left or right commands translation along the Y-axis. Moving the tee-handle inward or outward commands translation along the Z-axis. Moving the tee-handle upward or downward commands translation along the X-axis, using the RCS thrusters when the select lever is in the down position. When the lever is in the up position, upward or downward movement of the controller increases or decreases, respectively, the magnitude of descent engine thrust.

The controller is spring loaded to its neutral position in all axes when the lever is in jets position. When the lever is in the throttle position the Y and Z axes movements are spring loaded to the neutral position but the X-axis throttle commands will remain at the position set by the astronauts.

ATTITUDE AND TRANSLATION CONTROL ASSEMBLY

The attitude and translation control assembly controls LM attitude and translation. In the primary guidance path, attitude and translation commands are generated by the primary guidance computer and applied directly to jet drivers within the assembly. In the abort guidance path, the attitude and translation control assembly receives translation commands from the thrust/translation controller assembly, rate-damping signals from the rate gyro assembly, and attitude rate commands and pulse commands from the attitude controller assembly.

The assembly combines attitude and translation commands in its logic network to select the proper thruster to be fired for the desired combination of translation and rotation.

RATE GYRO ASSEMBLY

The rate gyro assembly consists of three single-degree-of-freedom rate gyros mounted so that they sense vehicle roll, pitch, and yaw rates. Each rate gyro senses a rate of turn about its input axis, which is perpendicular to the spin and output axes. The rate of turn is dependent on the gimbal position of the gyro. In abort guidance control, pickoff voltages are routed to the attitude and translation control assembly for rate damping.

DESCENT ENGINE CONTROL ASSEMBLY

The descent engine control assembly accepts engine-on and engine-off commands from the S&C control assemblies, throttle commands from the primary guidance computer and the thrust/translation controller assembly, and trim commands from the primary guidance computer or the attitude and translation control assembly. Demodulators, comparators, and relay logic circuits convert these inputs to the required descent engine commands. The assembly applies throttle and engine control commands to the descent engine and routes trim commands to the gimbal drive actuators.

Under normal operating conditions with primary guidance in control, the descent engine is manually selected and armed by an astronaut action. The descent engine control assembly responds by routing, through relay logic, 28 volts dc to the actuator isolation solenoids of the descent engine. Once the engine is armed, the assembly receives an automatic descent engine-on command from the primary guidance computer or a descent engine-on command initiated by the Commander pressing the start pushbutton. When the engine is fired, the descent engine control switching and logic latch the engine in the on position until an automatic or manual off command is received by the assembly. When the measured change in velocity reaches a predetermined value, the primary guidance computer generates a descent engine-off command. Manual engine commands are generated by the astronauts and will override the automatic function.

The control assembly accepts manual and automatic throttle commands from the thrust/translation controller assembly and the primary guidance computer, respectively. Manual or automatic thrust control is selected by the astronaut. During manual throttle control, computer throttle commands are interrupted and only manual commands are accepted by the assembly. The astronauts can monitor the response to their manual commands on the thrust indicator. Manual throttle commands consist of 800-Hz a-c voltages which are proportional to X-axis displacement of the thrust/translation control assembly. The active controller always provides at least a 10% command. These commands drive a nonlinear circuit to provide the desired thrust level. At an approximately 60% thrust the nonlinear region of the thrust/translation controller assembly is reached; it is displaced to its hard stop (92.5% thrust) to prevent erratic descent engine operation.

Automatic throttle increase or decrease commands are generated by the primary guidance computer under program control. These are predetermined levels of thrust and can be overridden by the astronaut using his thrust controller. No provision is made for automatically throttling the engine, using the abort guidance computer. The automatic commands appear on two separate lines (throttle increase and throttle decrease) as 3,200-Hz pulse inputs to an integrating d-c counter (up-down counter). Each pulse corresponds to a 2.7-pound thrust increment.

During automatic throttle operation, computer-commanded thrust is summed with the output of the thrust/translation controller. When the thrust/translation controller is in its minimum position, the computer-commanded thrust is summed with the fixed 10% output of the controller. When an active controller is displaced from its minimum position, the amount of manual thrust commanded is summed with the computer-commanded thrust to produce the desired resultant. In this case, the controller overrides the computer's control of descent engine thrust. The total thrust commanded (automatic and/or manual) cannot exceed 92.5%. Automatic thrust commands derived by the computer are always 10% lower than required thrust to compensate for the fixed output of the thrust controller.

Two channels of electronics are provided to control the roll and pitch position of the descent engine thrust vector with respect to the vehicle's center of gravity. When the descent engine is firing, this trim control acts as a low-frequency stabilization system in parallel with the higher frequency RCS. Each channel is driven by either the primary guidance computer when the primary guidance mode is used; by the attitude and translation control assembly when the abort guidance mode is used.

In the primary guidance mode, the computer provides automatic trim control. When the computer determines the required descent engine trim, it provides a trim command to the descent engine control assembly, on a positive or negative trim line for the pitch or roll axis. The trim command is routed to a malfunction logic circuit and to a power-switching circuit, which applies 115-volt, 400-Hz power to the proper gimbal drive actuator. In the abort guidance mode, trim commands are provided by the descent engine control assembly, by using the analog trim signals generated in the pitch and roll error channels of the attitude and translation control assembly.

LANDING RADAR

ELECTRONICS ASSEMBLY

The electronics assembly comprises frequency trackers (one for each velocity beam), a range frequency tracker, velocity converter and computer, range computer, signal data converter, and data-good/no-good logic circuit.

ANTENNA ASSEMBLY

The assembly comprises four microwave mixers, four dual audio-frequency preamplifiers, two microwave transmitters, a frequency modulator, and an antenna pedestal tilt mechanism.

The antenna consists of six planar arrays: two for transmission and four for reception. They are mounted on the tilt mechanism, beneath the descent stage, and may be placed in one of two fixed positions.

RENDEZVOUS RADAR

ELECTRONICS ASSEMBLY

The electronics assembly comprises a receiver, frequency synthesizer, frequency tracker, range tracker, servo electronics, a signal data converter, self-test circuitry, and a power supply. The assembly furnishes crystal-controlled signals, which drive the antenna assembly transmitter; provides a reference for receiving and processing the return signal; and supplies signals for antenna positioning.

ANTENNA ASSEMBLY

The main portion of the rendezvous radar antenna is a 24-inch parabolic reflector. A 4.65-inch hyperbolic subreflector is supported by four converging struts. Before the radar is used, the antenna is manually released from its stowed position. The antenna pedestal and the base of the antenna assembly are mounted on the external structural members of the LM. The antenna pedestal includes rotating assemblies that contain radar components. The rotating assemblies are balanced about a shaft axis and a trunnion axis. The trunnion axis is perpendicular to, and intersects, the shaft axis. The antenna reflectors and the microwave and RF electronics components are assembled at the top of the trunnion axis. This assembly is counterbalanced by the trunnion-axis rotating components (gyroscopes, resolvers, and drive motors) mounted below the shaft axis. Both groups of components, mounted opposite each other on the trunnion axis, revolve about the shaft axis. This balanced arrangement requires less driving torque and reduces the overall antenna weight. The microwave, radiating, and gimbaling components, and other internally mounted components, have low-frequency flexible cables that connect the outboard antenna components to the inboard electronics assembly.

MAIN PROPULSION
QUICK REFERENCE DATA

DESCENT PROPULSION SECTION

Pressurization section

   Ambient helium tank
Volume	1 cubic foot
Initial filling weight of helium	1.12 pounds
Initial helium pressure and temperature	1,635 psia at +70° F
Proof pressure	2,333 psi

   Supercritical helium tank
Volume	5.9 cubic feet
Initial filling weight of helium	51.2 pounds
Initial helium fllling pressure and temperature	178 psia at -450° F
Nominal helium storage pressure and temperature	1.555 psia at -400° F
Maximum helium storage pressure and temperature	1,780 psia at -312° F
Density	8.7 pounds per cubic foot
Proof pressure	2,274 psi
Burst-disk rupture pressure	1,881 to 1,967 psi

   Helium filters absolute filtration   15 microns

   Helium pressure regulators
Outlet pressure	245 ± 3 psia
Normal operation flow rate range	0.52 to 5.5 pounds per minute
Nominal flow rate at full throttle	5.2 pounds per minute
Inlet pressure range	320 to 1,750 psia
Maximum lockup pressure	253 psia at inlet pressure of 400 to 1,750 psia
	255 psia at inlet pressure of 320 to 400 psia

   Relief valve assembly
Burst-disk rupture pressure	260 to 275 psi
Relief valve cracking pressure	260 psi
Fully open flow rate	10 pounds per minute
Minimum reseat pressure	254 psi

Propellant feed section

   Propellant tanks
Capacity (each tank)	67.3 cubic feet
Total fuel	7,513 pounds
Total oxidizer	11,993 pounds
Minimum ullage volume (each tank)	1,728 cubic inches
Usable fuel	7,492 pounds
Usable oxidizer	11,953 pounds
Nominal ullage pressure (at full throttle position)	235 psia
Nominal propellant temperature	+70° F
Propellant temperature range	+50° to +90° F
Proof pressure	360 psia
Propellant filters absolute filtration	60 microns

Engine assembly

Nominal engine thrust (full throttle)	9,900 pounds (FTP)
Minimum engine thrust (low stop)	1,280 pounds (12.2%)
Nominal combustion chamber pressure (FTP)	103.4 psia
Engine-gimbaling capability	+6° to -6° from center, along Y-axis and Z-axis
Propellant injection ratio (oxidizer to fuel)	1.6 to 1 (approximate)
Engine restart capability	20 times
Engine life	910 seconds or 17,510 pounds of propellant consumption
Approximate weight	348 pounds
Overall length	95 inches
Nozzle expansion area ratio	47.4 to 1
Nozzle exit diameter	63 inches

ASCENT PROPULSION SECTION

Pressurization section

Helium tanks

Volume (each tank)	3.35 cubic feet
Initial filling weight of helium (each tank)	6.5 pounds
Initial helium pressure and temperature	3,050 psia at +70° F
Maximum operating pressure of helium	3,500 psia at +160° F
Proof pressure	4,650 psia at +160° F

Helium filters absolute filtration	15 microns

Helium pressure regulator assemblies

Primary path outlet pressure

Upstream regulator	184 ± 4 psia
Downstream regulator	190 ± 4 psia

Secondary path outlet pressure

Upstream regulator	176 ± 4 psia
Downstream regulator	182 ± 4 psia
Maximum lockup pressure	203 psia
Maximum outlet flow rate (each regulator path)	5.5 pounds per minute
Inlet pressure range	400 to 3,500 psia
Nominal helium flow rate	1.45 pounds per minute

Relief valve assembly

    Burst-disk rupture pressure   226 to 250 psia

    Relief valve cracking pressure   245 psia

    Fully open flow rate   4 pounds per minute

    Reseat pressure   225 psia

Propellant feed section

    Propellant tanks

        Capacity (each tank)   36 cubic feet

        Total fuel   2,011 pounds

        Total oxidizer   3,217 pounds

        Minimum ullage volume (each tank)   0.5 cubic foot per tank at +90° F

        Usable fuel   2,001 pounds

        Usable oxidizer   3,190 pounds

        Propellant temperature range   +50° to +90° F

        Nominal propellant temperature   +70° F

        Nominal ullage pressure   184 psia

        Proof pressure   333 psia

    Propellant filters absolute filtration   200 microns

    Engine assembly

        Nominal engine thrust   3,500 pounds

        Nominal combustion chamber pressure   120 psia

        Fuel flow rate   4.3 pounds per second

        Oxidizer flow rate   6.9 pounds per second

        Propellant injection ratio (oxidizer to fuel)   1.6 to 1

        Injector inlet pressure

            Steady-state operation   145 psia

            Engine start   185 to 203 psia

        Engine start to 90% of rated thrust   0.310 second

        Engine shutdown to 10% of rated thrust   0.200 second

        Nominal propellant temperature at injector inlet   +70° F

        Restart capability   35 times

        Engine life   460 seconds

        Approximate weight   200 pounds

        Overall length   47 inches

        Nozzle expansion area ratio   45.6 to 1

        Nozzle exit diameter   31 inches

The Main Propulsion Subsystem (MPS) consists of two separate, complete, and independent propulsion sections: the descent propulsion section and the ascent propulsion section. Each propulsion section performs a series of specific tasks during the lunar-landing mission. The descent propulsion section provides propulsion for the LM from the time it separates from the CSM until it lands on the lunar surface, the ascent propulsion section lifts the ascent stage off the lunar surface and boosts it into orbit. Both propulsion sections operate in conjunction with the Reaction Control Subsystem (RCS), which provides propulsion used mainly for precise attitude and translation maneuvers. The ascent propellant tanks are connected to the RCS to supplement its propellant supply during certain mission phases. If a mission abort becomes necessary during the descent trajectory, the ascent or descent engine can be used to return to a rendezvous orbit with the CSM. The choice of engines

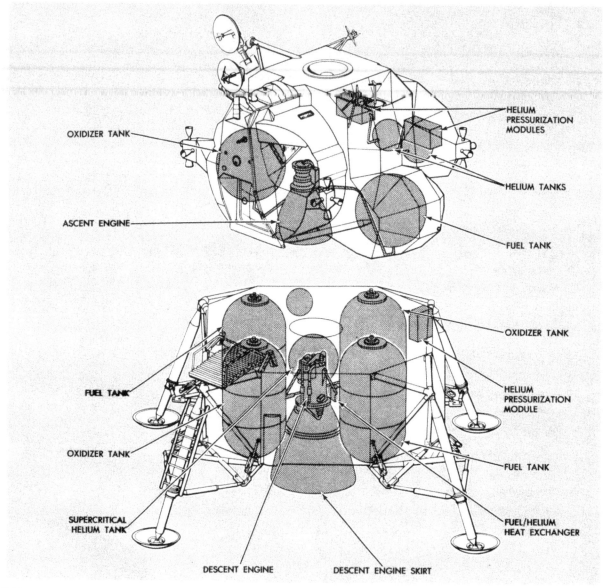

Major Main Propulsion Equipment Location

R-75

depends on the cause for abort, the amount of propellant remaining in the descent stage, and the length of time that the descent engine had been firing.

Each propulsion section consists of a liquid-propellant, pressure-fed rocket engine and propellant storage, pressurization, and feed components. For reliability, many vital components in each section are redundant. In both propulsion sections, pressurized helium forces the hypergolic propellants from the tanks to the engine injector. Both engine assemblies have control valves that start and stop a metered propellant flow to the combusion chamber upon command, trim orifices, an injector that determines the spray pattern of the propellants as they enter the combustion chamber, and a combustion chamber, where the propellants meet and ignite. The gases produced by combustion pass through a throat area into the engine nozzle, where they expand at an extremely high velocity before being ejected. The momentum of the exhaust gases produces the reactive force that propels the vehicle.

The more complicated tasks required of the descent propulsion section — such as propelling the entire LM and hovering over the lunar surface while the astronauts select a landing site — dictate that the descent propulsion section be the larger and more sophisticated of the two propulsion sections. It has a propellant supply that is more than three times that of the ascent propulsion section. The descent engine is almost twice as large as the ascent engine, produces more thrust (almost 10,000 pounds at full throttle), is throttleable for thrust control, and is gimbaled (can be tilted) for thrust vector control. The ascent engine, which cannot be tilted, delivers a fixed thrust of 3,500 pounds, sufficient to launch the ascent stage from the lunar surface and place it into a predetermined orbit.

The primary characteristics demanded of the LM propellants are high performance per weight; storability over long periods without undue vaporization or pressure buildup; hypergolicity for easy, closely spaced engine starts; no shock sensitivity; freezing and boiling points within controllable extremes; and chemical stability. The ascent and descent propulsion sections, as well as the RCS, use identical fuel/oxidizer combinations. In the ascent and descent propulsion sections, the injection ratio of oxidizer to fuel is approximately 1.6 to 1, by weight.

The fuel is a blend of hydrazine (N_2H_4) and unsymmetrical dimethylhydrazine (UDMH), commercially known as Aerozine 50. The proportions, by weight, are approximately 50% hydrazine, and 50% dimethylhydrazine.

The oxidizer is nitrogen tetroxide (N_2O_4). It has a minimum purity of 99.5% and a maximum water content of 0.1%.

The astronauts monitor the performance and status of the MPS with their panel-mounted pressure, temperature, and quantity indicators; talkbacks (flags indicating open or closed position of vital valves); and caution and warning annunciators (placarded lights that go on when specific out-of-tolerance conditions occur). These data, originating at sensors and position switches in the MPS, are processed in the Instrumentation Subsystem, and are simultaneously displayed to the astronauts in the LM cabin and transmitted to mission controllers through MSFN via the Communications Subsystem. The MPS obtains 28-volt d-c and 115-volt a-c primary power from the Electrical Power Subsystem.

Before starting either engine, the propellants must be settled to the bottom of the tanks. Under weightless conditions, this requires an ullage maneuver; that is, the LM must be moved in the +X, or upward, direction. To perform this maneuver, an astronaut or the automatic guidance equipment operates the downward-firing thrusters of the RCS.

The MPS is operated by the Guidance, Navigation, and Control Subsystem (GN&CS), which issues automatic (and processes manually initiated) on and off commands to the descent or ascent engine. The GN&CS also furnishes gimbal-drive and thrust-level commands to the descent propulsion section.

DESCENT ENGINE OPERATION AND CONTROL

After initial pressurization of the descent propulsion section, the descent engine start requires two separate and distinct operations: arming and firing. Engine arming is performed by the astronauts; engine firing can be performed by the astronauts, or it can be automatically initiated by the LM guidance computer. When the astronauts set a switch to arm the descent engine, power is simultaneously routed to open the actuator isolation valves in the descent engine, enable the instrumentation circuits in the descent propulsion section, and issue a command to the throttling controls to start the descent engine at the required 10% thrust level. The LM guidance computer and the abort guidance section receive an engine-armed status signal. This signal enables an automatic engine-on program in the GN&CS, resulting in a descent engine start. A manual start is accomplished when the Commander pushes his engine-start pushbutton. (Either astronaut can stop the engine because separate engine-stop pushbuttons are provided at both flight stations.)

The normal start profile for all descent engine starts must be at 10% throttle setting. Because the thrust vector at engine start may not be directed through the LM center of gravity, a low-thrust start (10%) will permit corrective gimbaling. If the engine is started at high thrust, RCS propellants must be used to stabilize the LM.

The astronauts can, with panel controls, select automatic or manual throttle control modes and Commander or LM Pilot thrust/translation controller authority, and can override automatic engine operation. Redundant circuits, under astronaut control, ensure descent engine operation if prime control circuits fail.

Signals from the GN&CS automatically control descent engine gimbal trim a maximum of 6° from the center position in the Y- and Z-axes to compensate for center-of-gravity offsets during descent engine firing. This ensures that the thrust vector passes through the LM center of gravity. The astronauts can control the gimbaling only to the extent that they can interrupt the tilt capability of the descent engine which they would do if a caution light indicates that the gimbal drive actuators are not following the gimbal commands.

DESCENT PROPULSION SECTION FUNCTIONAL DESCRIPTION

The descent propulsion section consists of an ambient and supercritical helium tank with associated helium pressurization components; two fuel and two oxidizer tanks with associated feed components; and a pressure-fed, ablative, throttleable rocket engine. The engine can be shut down and restarted as required by the mission. At the full-throttle position, the engine develops a nominal thrust of 9,900 pounds; it can also be operated within a range of 1,280 to 6,400 pounds of thrust. Functionally, the descent propulsion section can be subdivided into a pressurization section, a propellant feed section, and an engine assembly.

PRESSURIZATION SECTION

Before earth launch, all the LM propellant tanks are only partly pressurized (less than 230 psia), so that the tanks will be maintained within a safe pressure level under the temperature changes experienced during launch and earth orbit. Before initial engine start, the ullage space in each propellant tank requires additional pressurization. This initial pressurization is accomplished with a relatively small amount of helium stored at ambient temperature and at an intermediate pressure. To open the path from the ambient helium tank to the propellant tanks, the astronauts fire three explosive valves: an ambient helium isolation valve and the two propellant compatibility valves that prevent backflow of propellant vapors from degrading upstream components. After flowing through a filter, the ambient helium enters a pressure regulator which reduces the helium pressure to approximately 245 psi. The regulated helium then enters parallel paths which lead through quadruple check valves into the propellant tanks. The quadruple check valves, consisting of four valves in a series-parallel arrangement, permit flow in one direction only. This protects upstream components against corrosive propellant vapors and prevents hypergolic action due to backflow from the propellant tanks.

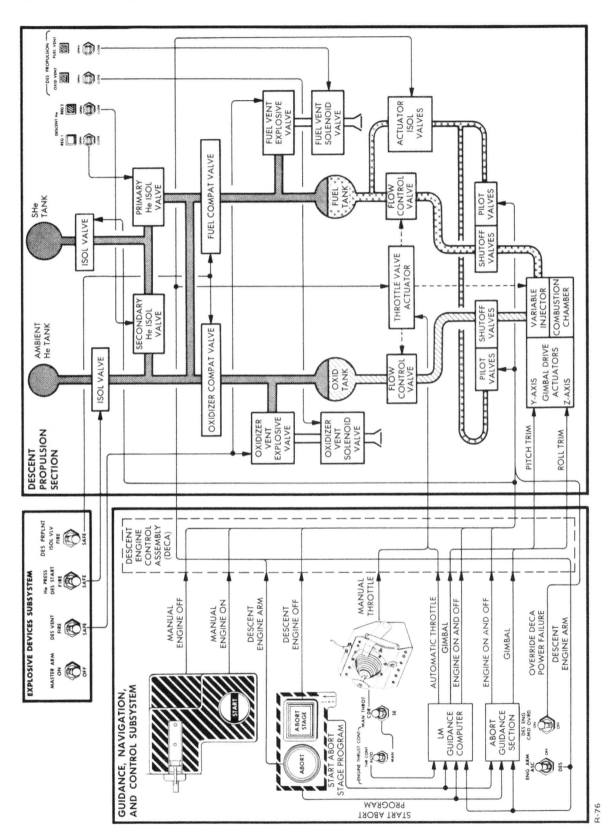

Descent Propulsion Control Diagram

SEQUENCE OF DESCENT ENGINE FIRING:

- PROPELLANT TANK PRESSURIZATION
  (INITIAL FIRING ONLY)
  PROPELLANT COMPATIBILITY
  EXPLOSIVE VALVES OPEN.
  AMBIENT He ISOLATION EXPLOSIVE
  VALVE OPENS.
- ENGINE ARMING
  ACTUATOR ISOLATION VALVES OPEN.
  THROTTLE VALVE ACTUATOR CIRCUITS
  ARE ENABLED.
- ENGINE FIRING
  PILOT VALVES.
  PROPELLANT SHUTOFF VALVES OPEN.
  SUPERCRITICAL He ISOLATION VALVE
  OPENS (INITIAL FIRING ONLY).

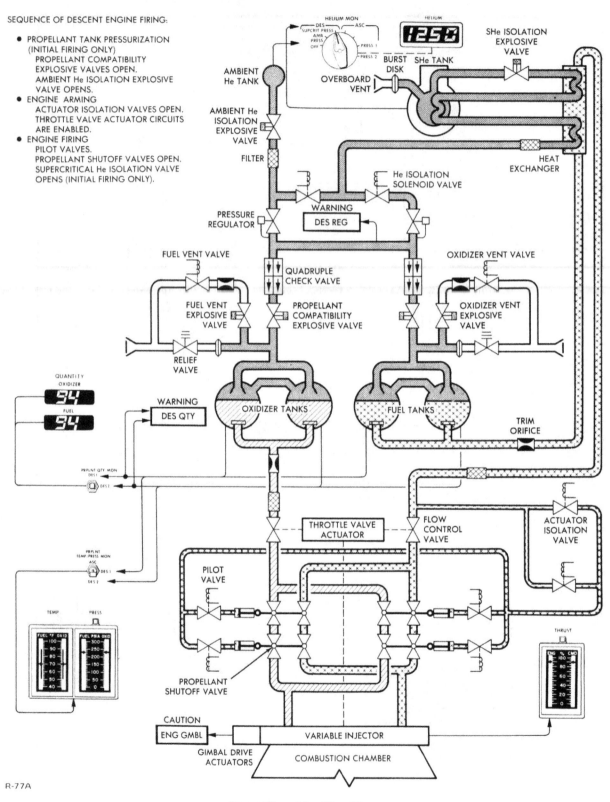

R-77A

Descent Propulsion Flow Diagram

After initial pressurization, supercritical helium is used to pressurize the propellants The supercritical helium tank is isolated by an explosive valve, which is automatically fired 1.3 seconds after the descent engine is started. The time delay prevents the supercritical helium from entering the fuel/helium heat exchanger until propellant flow is established so that the fuel cannot freeze in the heat exchanger. After the explosive valve opens, the supercritical helium enters the two-pass fuel/helium heat exchanger where it is slightly warmed by the fuel. The helium then flows back into a heat exchanger in the supercritical helium tank where it increases the temperature of the supercritical helium in the tank, causing a pressure rise and ensuring continuous expulsion of helium throughout the entire period of operation. Finally, the helium flows through the second loop of the fuel/helium heat exchanger where it is heated to operational temperature before it is regulated and routed to the propellent tanks.

The system that reduces the helium pressure consists of two parallel, redundant regulators. If one pressure regulator fails, the astronauts close the malfunctioning line and open the redundant line, to restore normal propellant tank pressurization.

Each propellant tank is protected against overpressurization by a relief valve, which opens at approximately 260 psia and reseats after overpressurization is relieved. A thrust neutralizer prevents the gas from generating unidirectional thrust. Each relief valve is paralleled by two series-connected vent valves, which are operated by panel switches. After landing, the astronauts relieve pressure buildup in the tanks, caused by rising temperatures, to prevent uncontrolled venting through the relief valves. The fuel and oxidizer fumes are vented separately; supercritical helium is vented at the same time.

PROPELLANT FEED SECTION

The descent section propellant supply is contained in two fuel tanks and two oxidizer tanks. Each pair of like propellant tanks is manifolded into a common delivery line.

Pressurized helium, acting on the surface of the propellant, forces the fuel and oxidizer into the delivery lines through a propellant retention device that maintains the propellant in the lines during negative-g acceleration. The oxidizer is piped directly to the engine assembly; the fuel circulates through the fuel/helium heat exchanger before it is routed to the engine assembly. Each delivery line contains a trim orifice and a woven, stainless-steel-wire-mesh filter. The trim orifices provide engine inlet pressure of approximately 222 psia at full throttle position. The filters prevent debris, originating at the explosive valves or in the propellant tanks, from contaminating downstream components.

ENGINE ASSEMBLY

The descent engine is mounted in the center compartment of the descent stage cruciform. Fuel and oxidizer entering the engine assembly are routed through flow control valves to the propellant shutoff valves. A total of eight propellant shutoff valves are used; they are arranged in series-parallel redundancy, four in the fuel line and four in the oxidizer line. The series redundancy ensures engine shutoff, should one valve fail to close. The parallel redundancy ensures engine start, should one valve fail to open.

To prevent rough engine starts, the engine is designed to allow the oxidizer to reach the injector first. The propellants are then injected into the combustion chamber, where hypergolic action occurs.

The propellant shutoff valves are actuator operated. The actuation line branches off the main fuel line at the engine inlet and passes through the parallel-redundant actuator isolation valves to four solenoid-operated pilot valves. From the pilot valves, the fuel enters the hydraulically operated actuators, which open the propellant shutoff valves. The actuator pistons are connected to rack-and-pinion linkages that rotate the balls of the shutoff valves 90° to the open position. The

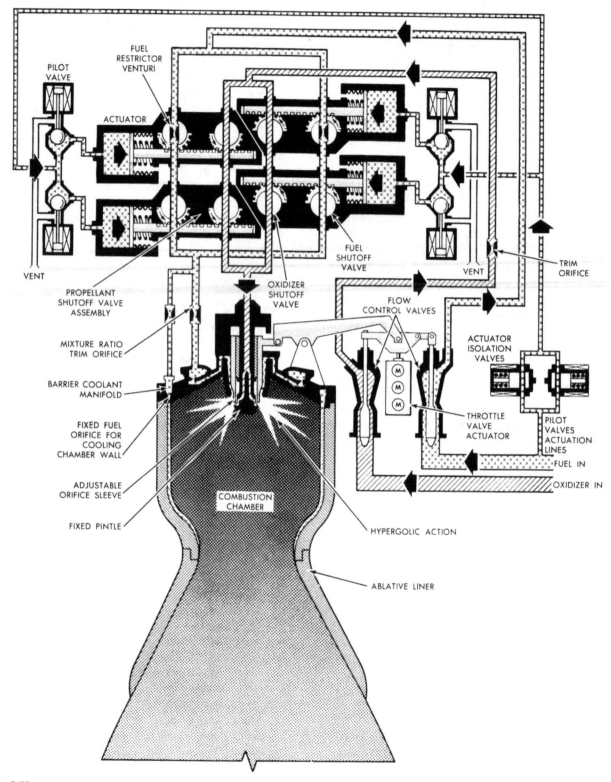

Descent Engine Flow Diagram

R-78

actuator isolation valves open when the astronauts arm the descent engine. When an engine-on command is initiated, the four pilot valves open simultaneously, permitting the actuation fuel to open the propellant shutoff ball valves, thus routing fuel and oxidizer to the combustion chamber.

The flow control valves, in conjunction with the adjustable orifice sleeve in the injector, control the descent engine thrust. At full throttle, and during the momentary transition from full throttle to the 65% range, throttling takes place primarily in the injector and, to a lesser degree, in the flow control valves. Below the 65% thrust level, the propellant-metering function is entirely controlled by the flow control valves. The flow control valves and

the injector sleeve are adjusted simultaneously by a mechanical linkage. Throttling is controlled by the throttle valve actuator, which positions the linkage in response to electrical input signals.

The fuel and oxidizer are injected into the combustion chamber at velocities and angles compatible with variations in weight flow. The fuel is emitted in the form of a thin cylindrical sheet; the oxidizer sprays break up the fuel stream and establish the injection pattern at all thrust settings. Some fuel is tapped off upstream of the injector and is routed through a trim orifice into the barrier coolant manifold. From here, it is sprayed against the combustion chamber wall through fixed orifices, maintaining the chamber wall at an acceptable temperature.

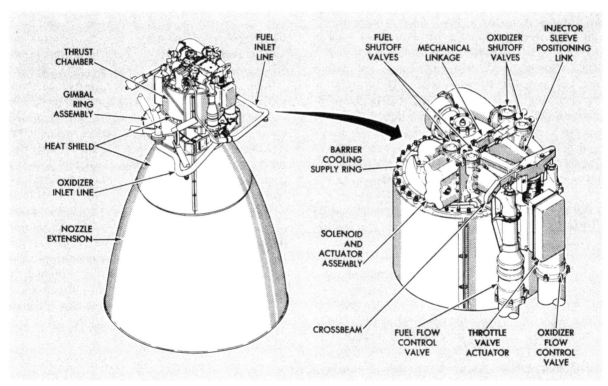

R-79

Descent Engine and Head End Assembly

DESCENT PROPULSION SECTION EQUIPMENT

SUPERCRITICAL HELIUM TANK

Supercritical helium is stored at a density approximately eight times that of ambient helium. Because heat transfer from the outside to the inside of the cryogenic storage vessel causes a gradual increase in pressure (approximately 10 psi per hour maximum), the initial loading pressure is planned so that the supercritical helium will be maintained within a safe pressure/time envelope throughout the mission.

The supercritical helium tank is double walled; it consists of an inner spherical tank and an outer jacket. The void between the tank and the jacket is filled with aluminized mylar insulation and evacuated to minimize ambient heat transfer into the tank. The vessel has fill and vent ports, a burst disk assembly, and an internal helium/helium heat exchanger. The inner tank is initially vented and loaded with cryogenic liquid helium at approximately 8° R (–452° F) at a pressure of 14.7 psia. The cryogenic liquid becomes supercritical helium when the fill sequence is completed by closing the vent and introducing a high-pressure head of gaseous helium. As the high-pressure, low-temperature gas is introduced, the density and pressure of the cryogenic liquid helium are increased. At the end of pressurization, the density of the stored supercritical helium is approximately 8.7 pounds per cubic foot and the final pressure is approximately 178 psi.

The burst disk assembly prevents hazardous overpressurization within the vessel. It consists of two burst disks in series, with a normally open, low-pressure vent valve between the disks. The burst disks are identical; they burst at a pressure between 1,881 and 1,967 psid to vent the entire supercritical helium supply overboard. A thrust neutralizer at the outlet of the downstream burst disk diverts the escaping gas into opposite directions to prevent unidirectional thrust generation. The vent valve prevents low-pressure buildup between the burst disks if the upstream burst disk leaks slightly. The valve is open at pressures below 150 psia; it closes when the pressure exceeds 150 psia.

FUEL/HELIUM HEAT EXCHANGER

Fuel is routed directly from the fuel tanks to the two-pass fuel/helium heat exchanger, where heat from the fuel is transferred to the supercritical helium. The helium reaches operating temperature after flowing through the second heat exchanger passage. The fuel/helium heat exchanger is of finned tube construction; the first and second helium passages are in parallel crossflow with respect to the fuel. Helium flows in the tubes and fuel flows in the outer shell across the bundle of staggered, straight tubes.

PROPELLANT STORAGE TANKS

The propellant supply is contained in four cylindrical, spherical-ended titanium tanks of identical size and construction. Two tanks contain fuel; the other two, oxidizer. Each pair of tanks containing like propellants is interconnected at the top and all propellant lines downstream of the tanks contain trim orifices, to ensure balanced propellant flow. A diffuser at the helium inlet port (top) of each tank distributes the pressurizing helium uniformly into the tank. An antivortex device in the form of a series of vanes, at each tank outlet, prevents the propellant from swirling into the outlet port, thus precluding inadvertent helium ingestion into the engine. Each tank outlet also has a propellant retention device (negative-g can) that permits unrestricted propellant flow from the tank under normal pressurization, but blocks reverse propellant flow (from the outlet line back into the tank) under zero-g or negative-g conditions. This arrangement ensures that helium does not enter the propellant outlet line as a result of a negative-g or zero-g condition or propellant vortexing; it eliminates the possibility of engine malfunction due to helium ingestion.

PROPELLANT QUANTITY GAGING SYSTEM

The propellant quantity gaging system enables the astronauts to monitor the quantity of propellants remaining in the four descent tanks. The propellant quantity gaging system consists of four quantity-sensing probes with low-level sensors (one for each tank), a control unit, two quantity indicators that display remaining fuel and oxidizer quantities, a switch that permits the astronauts to select a set of tanks (one fuel and one oxidizer) to be monitored, and a descent propellant quantity low-level warning light. The low-level sensors provide a discrete signal to cause the warning light to go on when the propellant level in any tank is down to 9.4 inches (equivalent to 5.6% propellant remaining). When this warning light goes on, the quantity of propellant remaining is sufficient for only 2 minutes of engine burn at hover thrust (approximately 25%).

PROPELLANT SHUTOFF VALVE ASSEMBLIES

Each of the four propellant shutoff valve assemblies consists of a fuel shutoff valve, an oxidizer shutoff valve, a pilot valve, and a shutoff valve actuator. The shutoff valve actuator and the fuel shutoff valve are in a common housing. The four solenoid-operated pilot valves control the fuel that is used as actuation fluid to open the fuel shutoff valves. The oxidizer shutoff valve is actuated by a mechanical linkage driven from the fuel shutoff valve. When the pilot valves are opened, the actuation fluid flow (at approximately 110 psia) acts against the spring-loaded actuator plunger, opening the shutoff valves. When the engine-firing signal is removed, the pilot valves close and seal off the actuation fluid. The propellant shutoff valves are closed by the return action of the actuator piston springs, which expels the fuel entrapped in the cylinders and valve passages through the pilot valve vent port.

The propellant shutoff valves are ball valves. The ball element operates against a spring-loaded soft seat to ensure positive sealing when the valve is closed. The individual valves are rotated by a rack-and-pinion-gear arrangement, which translates the linear displacement of the pistons in the shutoff valve actuators.

THROTTLE VALVE ACTUATOR

The throttle valve actuator is a linear-motion electromechanical servoactuator which moves the throttle linkage in response to an electrical input command. Moving the throttle linkage simultaneously changes the position of the flow control valve pintles and the injector sleeve, thereby varying the amount of fuel and oxidizer metered into the engine and changing the magnitude of engine thrust. The throttle valve actuator is located between the fuel and oxidizer flow control valves; its housing is rigidly attached to the engine head end and its output shaft is attached to the throttle linkage.

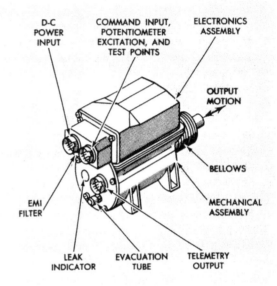

R-80

Throttle Valve Actuator

The actuator is controlled by three redundant electronic channels, which power three d-c torque motors. The motor shafts supply the input to a ball screw, which converts rotary motion to the linear motion of the throttle valve actuator output shaft. All mechanical moving parts of the actuator are within a hermetically sealed portion of the unit, pressurized to 0.25 psia with a 9 to 1 mixture of nitrogen and helium. A leak indicator in the cover provides visual evidence of loss of vacuum within the unit. Five potentiometers are ganged to the torque motor shaft through a single-stage planetary reduction gear. Three of these potentiometers supply position feedback information to the three motor amplifier channels, one to each channel. The other two potentiometers provide throttle actuator shaft position data for telemetry to MSFN. The redundancy within the throttle valve actuator ensures that failure of any electrical component will not cause the actuator to fail. The throttle valve actuator also provides a fail-safe system in the event selective malfunctions external to the throttle valve actuator occur. If either the primary 28-volt d-c power or the command voltage is lost, the throttle valve actuator causes the descent engine to thrust automatically at full throttle.

FLOW CONTROL VALVES

The oxidizer and fuel flow control valves are on the side of the engine, immediately downstream of the propellant inlet lines. They are secured to the throttle valve actuator mounting bracket. The flow control valve pintle assemblies are mechanically linked to the throttle valve actuator by a crossbeam.

The flow control valves are nonredundant cavitating venturis with movable pintle sleeves. Engine throttling is initiated by an electrical signal to the throttle valve actuator, commanding an increase or decrease in engine thrust. Operation of the throttle valve actuator changes the position of the pintles in the flow control valves. This axial movement of the pintles decreases or increases the pintle flow areas to control propellant flow rate and thrust. Below an approximate 70% thrust setting, flow through the valves cavitates, and hydraulically uncouples the propellant transfer system (and thereby, the flow rate) from variations in combustion chamber pressure. In the throttling range between 65% and 92.5% thrust, operation of the cavitating venturis of the flow control valves becomes unpredictable and may cause an improper fuel-oxidizer mixture ratio, which will result in excessive engine erosion and early combustion chamber burn-through.

VARIABLE-AREA INJECTOR

The variable-area injector consists of a pintle assembly, drive assembly, and manifold assembly. The pintle assembly introduces the propellant uniformly into the combustion chamber. The drive assembly has a twofold function: first, it serves as a passage for conducting the oxidizer into the pintle assembly; second, it contains the

bearing and sealing components that permit accurate positioning of the injector sleeve. The injector sleeve varies the injection area so that near-optimum injector pressure drops and propellant velocities are maintained at each thrust level. The primary function of the manifold assembly is to distribute the fuel uniformly around the outer surface of the sleeve. Fuel enters the manifold assembly at two locations and is passed through a series of distribution plates near the outer diameter of the assembly.

At the center of the manifold, the fuel passes through a series of holes before it is admitted into a narrow passage formed by the manifold body and a faceplate. The passage smoothes out gross fuel discontinuities and assists in cooling the injector face. The fuel then passes onto the outer surface of the sleeve, past a fuel-metering lip. The fuel is injected in the form of a hollow cylinder so that it reaches the impingement zone with a uniform circumferential velocity profile and without atomizing, at all flow rates. The oxidizer is injected through a double-slotted sleeve so that it forms a large number of radial filaments. Each filament partially penetrates the fuel cylinder and is enfolded by fuel in such a way that little separation of oxidizer and fuel can occur. For given propellant densities, overall mixture ratio, and injector geometry, there is a range of propellant injection velocity ratios that result in maximum mixture ratio uniformity throughout the resultant expanding propellant spray. When they occur, the liquid-phase reactions generate gas and vapor that atomize and distribute the remaining liquid oxidizer and fuel uniformly in all directions, resulting in high combustion efficiency.

COMBUSTION CHAMBER AND NOZZLE EXTENSION

The combustion chamber consists of an ablative-cooled chamber section, nozzle throat, and nozzle divergent section. The ablative sections are enclosed in a continuous titanium shell and jacketed in a thermal blanket composed of aluminized nickel foil and glass wool. A seal prevents leakage between the combustion chamber and nozzle extension.

The nozzle extension is a radiation-cooled, crushable skirt; it can collapse a distance of 28 inches on lunar impact so as not to affect the stability of the LM. The nozzle extension is made of columbium coated with aluminide. It is attached to the combustion chamber case at a nozzle area ratio of 16 to 1 and extends to an exit area ratio of 54.0 to 1.

GIMBAL RING AND GIMBAL DRIVE ACTUATORS

The gimbal ring is located at the plane of the combustion chamber throat. It consists of a rectangular beam frame and four trunnion subassemblies. The gimbal drive actuators under control of the descent engine control assembly, tilt the descent engine in the gimbal ring along the pitch and roll axes so that the engine thrust vector goes through the LM center of gravity. One actuator controls the pitch gimbal; the other, the roll gimbal. The gimbal drive actuators consist of a single-phase motor, a feedback potentiometer, and associated mechanical devices. They can extend or retract 2 inches from the mid-position to tilt the descent engine a maximum of 6° along the Y-axis and Z-axis.

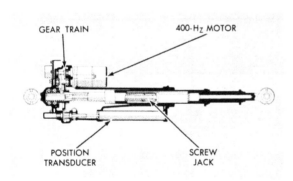

R-81

Gimbal Drive Actuator

ASCENT ENGINE OPERATION AND CONTROL

Shortly before initial ascent engine use, the astronauts fire explosive valves to pressurize the ascent propulsion section. The ascent engine, like the descent engine, requires manual arming before

it can be fired. When the astronauts arm the ascent engine, a shutoff command is sent to the descent engine. Then, enabling signals are sent to the ascent engine control circuitry to permit a manual or computer-initiated ascent engine start. For manual engine on and off commands, the astronauts push the same start and stop pushbuttons used for the descent engine. For automatic commands, the stabilization and control assemblies in the GN&CS provide sequential control of LM staging and ascent engine on and off commands. The initial ascent engine firing — whether for normal lift-off from the lunar surface or in-flight abort — is a fire-in-the-hole (FITH) operation; that is, the engine fires while the ascent and descent stages are still mated although no longer mechanically secured to each other. If, during the descent trajectory, an abort situation necessitates using the ascent engine to return to the CSM, the astronauts abort stage sequence. This results in an immediate descent engine shutdown followed by a time delay to ensure that the engine has stopped thrusting before staging occurs. The next command automatically pressurizes the ascent propellant tanks, after which the staging command is issued. This results in severing of hardware that secures the ascent stage to the descent stage and the interconnecting cables. The ascent engine fire command completes the abort stage sequence.

ASCENT PROPULSION SECTION FUNCTIONAL DESCRIPTION

The ascent propulsion section consists of a constant-thrust, pressure-fed rocket engine, one fuel and one oxidizer tank, two helium tanks, and associated propellant feed and helium pressurization components. The engine develops 3,500 pounds of thrust in a vacuum, it can be shut down and restarted, as required by the mission. Like the descent propulsion section, the ascent propulsion section can functionally be subdivided into a pressurization section, a propellant feed section, and an engine assembly.

PRESSURIZATION SECTION

Before initial ascent engine start, the propellant tanks must be fully pressurized with gaseous helium. This helium is stored in two identical

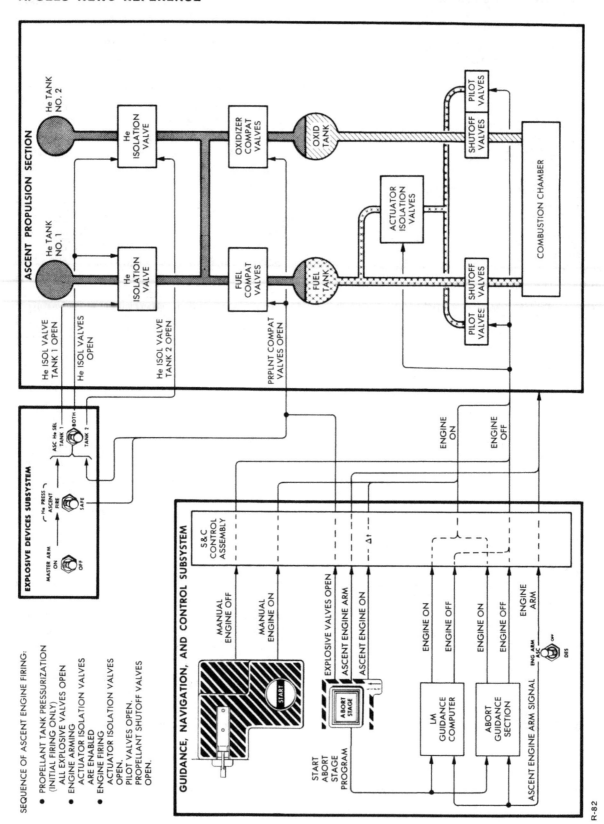

Ascent Propulsion Control Diagram

R-82

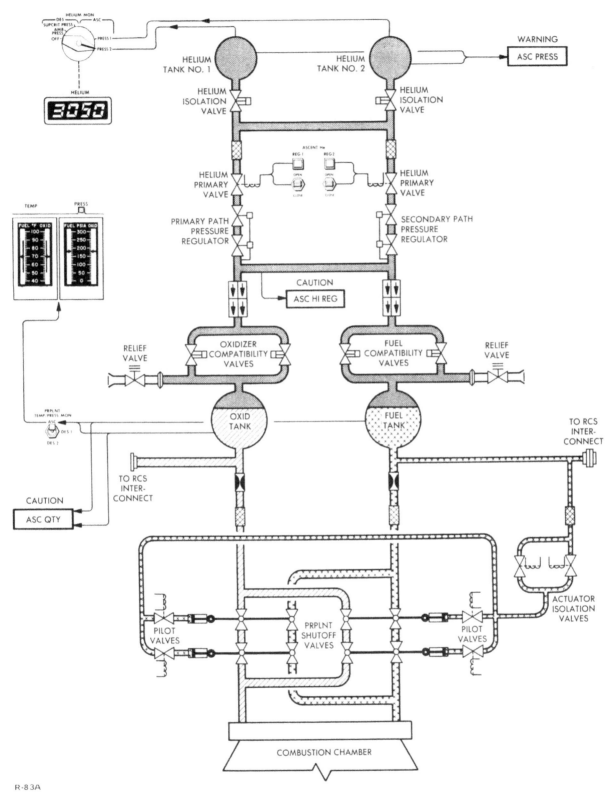

Ascent Propulsion Flow Diagram

R-83A

tanks at a nominal pressure of 3,050 psia at a temperature of +70° F. An explosive valve at the outlet of each helium tank prevents the helium from leaving the tanks until shortly before initial ascent engine use. To open the helium paths to the propellant tanks, the astronauts normally fire six explosive valves simultaneously: two helium isolation valves and four propellant compatibility valves (two connected in parallel for redundancy in each pressurization path). Before firing the explosive valves, the astronauts check the pressure in each helium tank. If one tank provides an unusually low reading (indicating leakage), they can exclude the appropriate helium isolation explosive valve from the fire command. This will isolate the faulty tank from the pressurization system and will prevent helium loss through the leaking tank via the helium interconnect line.

Downstream of the interconnect line, the helium flows into the primary and secondary regulating paths, each containing a filter, a normally open solenoid valve and two series-connected pressure regulators. Two downstream regulators are set to a slightly higher output pressure than the upstream regulators; the regulator pair in the primary flow path produces a slightly higher output than the pair in the secondary (redundant) flow path. This arrangement causes lockup of the regulators in the redundant flow path after the propellant tanks are pressurized, while the upstream regulator in the primary flow path maintains the propellant tanks at their normal pressure of 184 psia. If either regulator in the primary flow path fails closed, the regulators in the redundant flow path pressurize the propellant tanks. If an upstream regulator fails open, control is obtained through the downstream regulator in the same flow path. If both regulators in the same flow path fail open, pressure in the helium manifold increases above the acceptable limit of 220 psia, causing a caution light to go on. This advises the astronauts that they must identify the failed-open regulators and close the helium isolation solenoid valve in the malfunctioning flow path so that normal pressure can be restored.

Downstream of the regulators, a manifold routes the helium into two flow paths: one path leads to the oxidizer tank; the other, to the fuel tank. A quadruple check valve assembly, a series-

parallel arrangement in each path, isolates the upstream components from corrosive propellant vapors. The check valves also safeguard against possible hypergolic action in the common manifold, resulting from mixing of propellants or fumes flowing back from the propellant tanks. Immediately upstream of the fuel and oxidizer tanks, each helium path contains a burst disk and relief valve assembly to protect the propellant tanks against overpressurization. This assembly vents pressure in excess of approximately 245 psia and reseals the flow path after overpressurization is relieved. A thrust neutralizer eliminates unidirectional thrust generated by the escaping gas.

PROPELLANT FEED SECTION

The ascent propulsion section has one oxidizer tank and one fuel tank. Transducers in each tank enable the astronauts to monitor propellant temperature and ullage pressure. A caution light, activated by a low-level sensor in each tank warn the astronauts when the propellant supply has diminished to an amount sufficient for only 10 seconds of engine operation.

Helium flows into the top of the propellant tanks, where diffusers uniformly distribute it throughout the ullage space. The outflow from each propellant tank divides into two paths. The primary path routes each propellant through a trim orifice and a filter to the propellant shutoff valves in the engine assembly. The trim orifice provides an engine inlet pressure of 170 psia for proper propellant use. The secondary path connects the ascent propellant supply to the RCS. This interconnection permits the RCS to burn ascent propellants, providing the ascent tanks are pressurized and the ascent or descent engine is operating when the RCS thrusters are fired. A line branches off the RCS interconnect fuel path and leads to two parallel actuator isolation solenoid valves. This line routes fuel to the engine pilot valves that actuate the propellant shutoff valves.

ENGINE ASSEMBLY

The ascent engine is installed in the midsection of the ascent stage; it is tilted so that its centerline is 1.5° from the X-axis, in the +Z-direction.

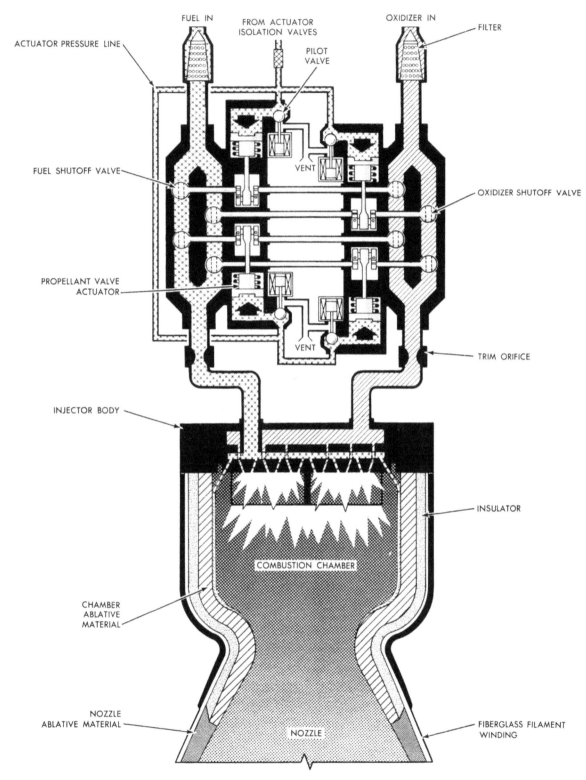

FUEL IN

FROM ACTUATOR
ISOLATION VALVES

OXIDIZER IN

FILTER

ACTUATOR PRESSURE LINE

PILOT
VALVE

FUEL SHUTOFF VALVE

VENT

OXIDIZER SHUTOFF VALVE

PROPELLANT VALVE
ACTUATOR

VENT

TRIM ORIFICE

INJECTOR BODY

INSULATOR

COMBUSTION CHAMBER

CHAMBER
ABLATIVE
MATERIAL

NOZZLE
ABLATIVE MATERIAL

NOZZLE

FIBERGLASS FILAMENT
WINDING

R-84

Ascent Engine Flow Diagram

Fuel and oxidizer entering the engine assembly are routed, through the filters, propellant shutoff valves, and trim orifices, to the injector. The propellants are injected into the combustion chamber, where the hypergolic ignition occurs. A separate fuel path leads from the actuator isolation valves to the pilot valves. The fuel in this line enters the actuators, which open the propellant shutoff valves.

Propellant flow into the combustion chamber is controlled by a valve package assembly, trim orifices, and the injector. The valve package assembly is similar to the propellant shutoff valve assemblies in the descent engine. The eight pro-

pellant shutoff valves are arranged in series-parallel redundant fuel-oxidizer pairs. Each pair is operated from a single crankshaft by its actuator.

When an engine-start command is received, the two actuator isolation valves and the four pilot valves open simultaneously. Fuel then flows through the actuator pressure line and the four pilot valves into the actuator chambers. Hydraulic pressure extends the actuator pistons, cranking the propellant shutoff valves 90° to the fully open position. The propellants now flow through the shutoff valves and a final set of trim orifices to the injector. The orifices trim the pressure differentials of the fuel and oxidizer to determine the mixture

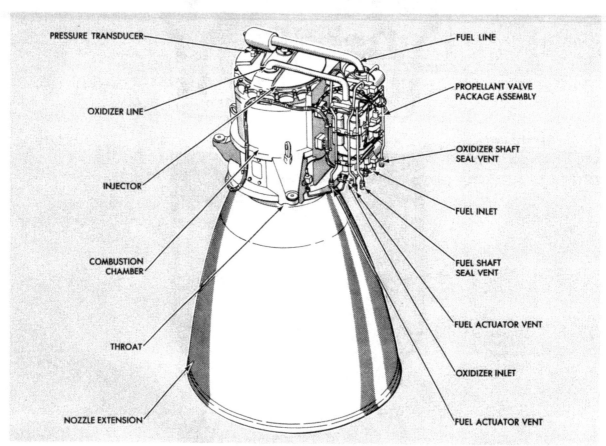

Ascent Engine Assembly

R-85

ratio of the propellants. The physical characteristics of the injector establish an oxidizer lead of approximately 50 milliseconds. This precludes the possibility of a fuel lead which could result in a rough engine start.

At engine shutdown, the actuator isolation valves are closed, preventing additional fuel from reaching the pilot valves. Simultaneously, the pilot valve solenoids are deenergized, opening the acutator ports to the overboard vents so that residual fuel in the actuators is vented into space. With the actuation fuel pressure removed, the actuator pistons are forced back by spring pressure, cranking the propellant shutoff valves to the closed position.

ASCENT PROPULSION SECTION EQUIPMENT

HELIUM PRESSURE REGULATOR ASSEMBLIES

Each helium pressure regulator assembly consists of two individual pressure regulators connected in series. The downstream regulator functions in the same manner as the upstream regulator; however, it is set to produce a higher outlet pressure so that it becomes a secondary unit that will only be in control if the upstream regulator (primary unit) fails open.

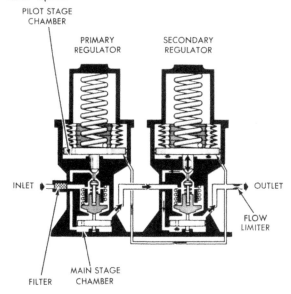

R-86

Helium Pressure Regulator Assembly

Each pressure regulator unit consists of a direct-sensing main stage and a pilot stage. The valve in the main stage is controlled by the valve in the pilot stage which senses small changes in the regulator outlet pressure and converts these changes to proportionally large changes in control pressure. A rise in outlet pressure decreases the pilot valve output, thereby reducing flow into the main stage chamber. An increase in the downstream demand causes a reduction in outlet pressure; this tends to open the pilot valve. The resultant increase in control pressure causes the main stage valve poppet to open, thus meeting the increased downstream demand.

A flow limiter at the outlet of the main stage valve of the secondary unit restricts maximum flow through the regulator assembly to 5.5 pounds of helium per minute, so that the propellant tanks are protected if the regulator fails open. The filter at the inlet of the primary unit prevents particles, which could cause excessive leakage at lockup, from reaching the regulator assembly.

PROPELLANT STORAGE TANKS

The propellant supply is contained in two spherical titanium tanks. The tanks are of identical size and construction. One tank contains fuel; the other, oxidizer. A helium diffuser at the inlet port of each tank distributes the pressurizing helium uniformly into the tank. An antivortex device (a cruciform at each tank outlet) prevents the propellant from swirling into the outlet port, precluding helium ingestion into the engine. Each tank outlet also has a propellant-retention device that permits unrestricted propellant flow from the tank under normal pressurization, but blocks reverse propellant flow (from the outlet line back into the tank) under zero-g or negative-g conditions. This arrangement ensures that helium does not enter the propellant outlet line while the engine is not firing; it eliminates the possibility of engine malfunction due to helium ingestion. A low-level sensor in each tank (approximately 4.4 inches above the tank bottom) supplies a discrete signal that causes a caution light to go on when the propellant remaining in either tank is sufficient for approximately 10 seconds of burn time (48 pounds of fuel, 69 pounds of oxidizer).

VALVE PACKAGE ASSEMBLY

At the propellant feed section/engine assembly interface, the oxidizer and fuel lines lead into the valve package assembly. The individual valves that make up the valve package assembly are in a series-parallel arrangement to provide redundant propellant flow paths and shutoff capability. The valve package assembly consists of eight propellant shutoff valves and four solenoid-operated pilot valve and actuator assemblies. Each valve assembly consists of one fuel shutoff valve and one oxidizer shutoff valve. These are ball valves that are operated by a common shaft, which is connected to its respective pilot valve and actuator assembly. Shaft seals and vented cavities prevent the propellants from coming into contact with each other. Separate overboard vent manifold assemblies drain the fuel and oxidizer that leaks past the valve seals, and the actuation fluid (fuel in the actuators when the pilot valves close), overboard. The eight shutoff valves open simultaneously to permit propellant flow to the engine while it is operating; they close simultaneously to terminate propellant flow at engine shutdown. The four nonlatching, solenoid-operated pilot valves control the actuation fluid (fuel).

INJECTOR ASSEMBLY

The injector assembly consists of the propellant inlet lines, a fuel manifold, a fuel reservoir chamber, an oxidizer manifold, and an injector orifice plate assembly. Because it takes longer to fill the fuel manifold and reservoir chamber assembly, the oxidizer reaches the combustion chamber approximately 50 milliseconds before the fuel, resulting in smooth engine starts. The injector orifice plate assembly is of the fixed-orifice type, which uses a baffle and a series of perimeter slots (acoustic cavities) for damping induced combustion disturbances. The baffle is Y-shaped, with a 120° angle between each blade. The baffle is cooled by the propellants, which subsequently enter the combustion chamber through orifices on the baffle blades. The injector face is divided into two combustion zones: primary and baffle. The primary zone uses impinging doublets (one fuel and one oxidizer), which are spaced in concentric radial rings on the injector face. The baffle zone (1.75 inches below the injector face) uses impinging doublets placed at an angle to the injector face radius. The combustion chamber wall is cooled by spraying fuel against it through canted orifices, spaced around the perimeter of the injector. The nominal temperature of the propellant is +70° F as it enters the injector; with the fuel temperature within 10° of the oxidizer temperature. The temperature range at engine start may be +40° to +500° F.

COMBUSTION CHAMBER ASSEMBLY AND NOZZLE EXTENSION

The combustion chamber assembly consists of the engine case and mount assembly and an ablative material (plastic) assembly, which includes the nozzle extension. The two assemblies are bonded and locked together to form an integral unit. The plastic assembly provides ablative cooling for the combustion chamber; it consists of the chamber ablative material, the chamber insulator, the nozzle extension ablative material, and a structural filament winding. The chamber ablative material extends from the injector to an expansion ratio of 4.6. The chamber insulator, between the ablative material and the case, maintains the chamber skin temperature within design requirements. The ablative material of the nozzle extension extends from the expansion ratio of 4.6 to 45.6 (exit plane) and provides ablative cooling in this region. The structural filament winding provides structural support for the plastic assembly and ties the chamber and nozzle extension sections together.

REACTION CONTROL

QUICK REFERENCE DATA

Pressurization section

    Helium tanks

    Unpressurized volume (each tank)   910 cubic inches

        Initial fill pressure and temperature $3,050 \pm 50$ psia at $+70^{\circ}$ F

        Initial filling weight of helium (each tank)   1.03 pounds

        Helium temperature range $+40^{\circ}$ to $+100^{\circ}$ F

        Proof pressure   4,650 psia

        Diameter   12.3 inches

    Helium filter absolute filtration   12 microns

    Primary pressure regulator

        Output 181 ± 3 psia

        Lockup pressure   188 psia (maximum)

    Secondary pressure regulator

        Output 185 ± 3 psia

        Lockup pressure   192 psia (maximum)

    Flow rate through pressure regulator assembly (single thruster operation)   0.036 pound per minute

    Relief valve assembly

        Venting pressure   232 psia

        Reseat pressure   212 psia (minimum)

        Burst-disk rupture pressure   220 psia

Propellant feed section

    Propellant tanks

        Working pressure   176 psia

        Proof pressure   333 psia

        Propellant pad pressure   50 psia

        Propellant storage temperature range $+40^{\circ}$ to 100° F

        Nominal temperature $+70^{\circ}$ F

        Diameter   12.5 inches

    Oxidizer tanks

        Volume (each tank)   2.38 cubic feet

        Ullage volume (each tank)   273.0 cubic inches

        Oxidizer flow rate to each thruster   0.240 pound per second

        Available oxidizer (each tank)   194.1 pounds

        Oxidizer loaded in each system (tank and manifold)   208.2 pounds

        Height   38 inches

    Fuel tanks

        Volume (each tank)   1.91 cubic feet

        Ullage volume (each tank)   158.5 cubic inches

        Fuel flow rate to each thruster   0.117 pounds per second

        Available fuel (each tank)   99.3 pounds (minimum)

        Fuel loaded in each system (tank and manifold)   107.4 pounds

        Height   32 inches

    Propellant filter absolute filtration   18 microns

    Ascent feed filter absolute filtration   25 microns

Thrust chamber assembly

    Engine thrust   100 pounds

    Engine life
        Total   1000 seconds
        Steady-state mode   500 seconds
        Pulse mode   500 seconds
        Restart capability   10,000 times

    Chamber-cooling method   Fuel-film cooling and radiation

    Combustion chamber pressure   96 psia

    Propellant injection ratio (oxidizer to fuel)   2.05 to 1

    Heaters
        Type   Resistance-wire element
        Operating power   28 volts dc
        Power consumption (each heater)   17.5 watts at 24 volts

    Oxidizer inlet pressure (steady state)   170±10 psia

    Fuel inlet pressure (steady state)   170±10 psia

    Approximate weight   5.25 pounds

    Overall length   13.5 inches

    Nozzle expansion area ratio   40 to 1

    Nozzle exit diameter   5.75 inches

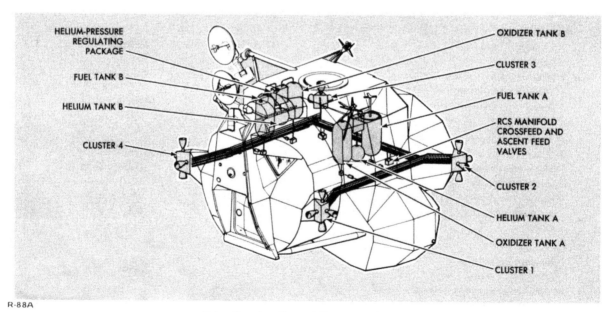

R-88A

Major Reaction Control Equipment Location

GRUMMAN

The Reaction Control Subsystem (RCS) provides thrust impulses that stabilize the LM during the descent and ascent trajectory and controls attitude and translation — movement of the LM about and along its three axes — during hover, landing, rendezvous, and docking maneuvers. The RCS also provides the thrust required to separate the LM from the CSM and the +X-axis acceleration (ullage maneuver) required to settle Main Propulsion Subsystem (MPS) propellants before a descent or ascent engine start. The RCS accomplishes its task during coasting periods or while the descent or ascent engine is firing; it operates in response to automatic control commands from the Guidance, Navigation, and Control Subsystem (GN&CS) or manual commands from the astronauts.

The 16 thrust chamber assemblies (thrusters) and the propellant and helium sections that comprise the RCS are located in or on the ascent stage. The propellants used in the RCS are identical with those used in the MPS. The fuel — Aerozine 50 — is a mixture of approximately 50% each of hydrazine and unsymmetrical dimethylhydrazine. The oxidizer is nitrogen tetroxide. The injection ratio of oxidizer to fuel is approximately 2 to 1. The propellants are hypergolic; that is, they ignite spontaneously when they come in contact with each other.

The thrusters are small rocket engines, each capable of delivering 100 pounds of thrust. They are arranged in clusters of four, mounted on four outriggers equally spaced around the ascent stage. In each cluster, two thrusters are mounted parallel to the LM X-axis, facing in opposite directions; the other two are spaced 90° apart, in a plane normal to the X-axis and parallel to the Y-axis and Z-axis.

The RCS is made up of two parallel, independent systems (A and B), which, under normal conditions, function together to provide complete attitude and translation control. Each system consists of eight thrusters, a helium pressurization section, and a propellant feed section. The two systems are interconnected by a normally closed crossfeed arrangement that enables the astronauts to operate all 16 thrusters from a single propellant supply. Complete attitude and translation control

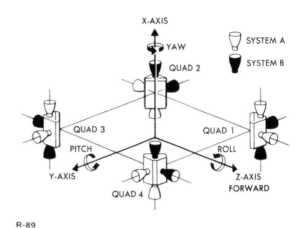

R-89

Thruster Arrangement

is therefore available even if one system's propellant supply is depleted or fails. Functioning alone, either RCS system can control the LM, although with slightly reduced efficiency. This capability is due to the distribution of the thrusters, because each cluster has two thrusters of each system located in a relatively different position.

In addition to the RCS propellant supply, the thrusters can use propellants from the ascent propulsion section. This method of feeding the thrusters, which requires the astronauts to open interconnect lines between the ascent tanks and RCS manifolds, is normally used only during periods of ascent engine thrusting. Use of ascent propulsion section propellants is intended to conserve RCS propellants, which may be needed during docking maneuvers.

The astronauts monitor performance and status of the RCS with their panel-mounted pressure, temperature, and quantity indicators; talkbacks (flags, that indicate open or closed position of certain valves); and caution and warning annunciators (placarded lights that go on when specific out-of-tolerance conditions occur). These data originate at sensors and position switches in the RCS, are processed in the Instrumentation Subsystem, and are simultaneously displayed to the astronauts in the LM cabin and transmitted to mission controllers through MSFN via the Communications Subsystem.

RCS Control Diagram

GRUMMAN

The 28-volt d-c and 115-volt a-c primary power required by the RCS is furnished by the Electrical Power Subsystem. Interconnect plumbing between the RCS thruster propellant manifolds and the ascent propulsion section tanks permit the RCS to use propellants from the Main Propulsion Subsystem (MPS) during certain phases of the mission.

Control of the RCS is provided by the GN&CS. Modes of operation, thruster selection, and firing duration are determined by the GN&CS.

FUNCTIONAL DESCRIPTION

THRUSTER SELECTION, OPERATION, AND CONTROL

The GN&CS provides commands that select thrusters and fire them for durations ranging from a short pulse to steady-state operation. The thrusters can be operated in an automatic mode, attitude-hold mode, or a manual override mode.

Normally, the RCS operates in the automatic mode; all navigation, guidance, stabilization, and steering functions are initiated and commanded by the LM guidance computer (primary guidance and navigation section) or the abort electronics assembly (abort guidance section).

The attitude-hold mode is a semiautomatic mode in which either astronaut can institute attitude and translation changes. When an astronaut displaces his attitude controller, an impulse proportional to the amount of displacement is routed to the computer, where it is used to perform steering calculations and to generate the appropriate thruster-on command. An input into the DSKY determines whether the computer commands an angular rate change proportional to attitude controller displacement, or a minimum impulse each time the controller is displaced. When the astronaut returns his attitude controller to the neutral (detent) position, the computer issues a command to maintain attitude. For a translation maneuver, either astronaut displaces his thrust/translation controller. This sends a discrete to the computer to issue a thruster-on command to selected thrusters. When this controller is returned to neutral, the thrusters cease to fire.

If the abort guidance section is in control, attitude errors are summed with the proportional rate commands from the attitude controller and a rate-damping signal from the rate gyro assembly. The abort guidance equipment uses this data to perform steering calculations, which result in specific thruster-on commands. The astronauts can select two or four X-axis thrusters for translation maneuvers, and they can inhibit the four upward-firing thrusters during the ascent thrust phase, thus conserving propellants. In the manual mode, the four-jet hardover maneuver, instituted when either astronaut displaces his attitude controller fully against the hard stop, fires four thrusters simultaneously, overriding any automatic commands.

For the MPS ullage maneuver, the astronauts select whether two or four downward-firing thrusters should be used. Depending on which guidance section is in control, the astronauts enter a DSKY input (primary) or use a 2-jet/4-jet selector switch (abort) to make their selection. Under manual control, a +X-translation pushbutton fires the four downward-firing thrusters continuously until the pushbutton is released. Firing two thrusters conserves RCS propellants; however, it takes longer to settle the MPS propellants.

RCS OPERATION

Functionally, the RCS can be subdivided into pressurization sections, propellant feed sections, and thruster sections. Because RCS systems A and B are identical, only one system is described.

Fuel and oxidizer are loaded into bladders within the propellant tanks and into the manifold plumbing that extends from the tanks through the normally open main shutoff valves up to the solenoid valves at each thruster pair. Before separation of the LM from the CSM, the astronauts set switches on the control panel to preheat the thrusters and fire explosive valves to pressurize the propellant tanks. Gaseous helium, reduced to a working pressure, enters the propellant tanks and forces the fuel and oxidizer to the thrusters. Here, the propellants are blocked by fuel and oxidizer valves that remain closed until a thruster-on command is issued. As the selected thruster receives the

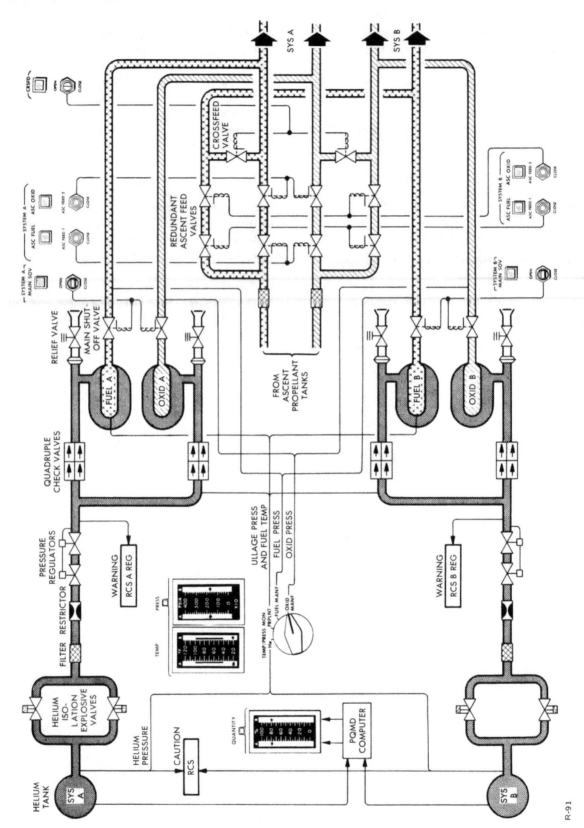

Helium Pressurization and Propellant Feed Sections Flow Diagram

R-91

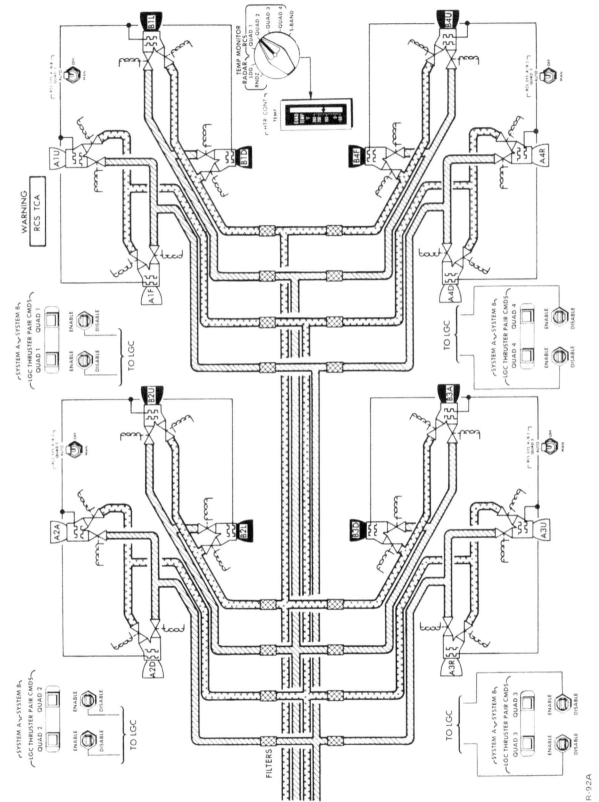

Propellant Lines and Thrusters Flow Diagram

R-92A

fire command, its fuel and oxidizer valves open to route the propellants through an injector into the combustion chamber, where they impinge and ignite by hypergolic action. The astronauts can disable malfunctioning thrusters by operating appropriate LGC thruster pair command switches on the control panel. When any of these switches is in the disable position, it issues a signal informing the LM guidance computer that the related thruster pair is disabled and that alternate thrusters must be selected. Talkbacks above each switch informs the astronauts of the status of related thruster pair.

PRESSURIZATION SECTION

The RCS propellants are pressurized with high-pressure gaseous helium, stored at ambient temperature. The helium tank outlet is sealed by parallel-connected, redundant helium isolation explosive valves that maintain the helium in the tank until the astronauts enter the LM and prepare the RCS for operation. When the explosive valves are fired, helium enters the pressurization line and flows through a filter. A restrictor orifice, downstream of the filter, dampens the initial helium surge.

Downstream of the restrictor, the flow path contains a pair of pressure regulators connected in series. The primary (upstream) regulator is set to reduce pressure to approximately 181 psia. The secondary (downstream) regulator is set for a slightly higher output (approximately 185 psia). In normal operation, the primary regulator is in control and provides proper propellant tank pressurization.

Downstream of the pressure regulators, a manifold divides the helium flow into two paths: one leads to the oxidizer tank; the other, to the fuel tank. Each flow has quadruple check valves that permit flow in one direction only, thus preventing backflow of propellant vapors if seepage occurs in the propellant tank bladders. A relief valve assembly protects each propellant tank against over-pressurization. If helium pressure builds up to 232 psia, the relief valve opens to relieve pressure by venting helium overboard. At 212 psia, the relief valve closes.

PROPELLANT FEED SECTION

Fuel and oxidizer are contained in flexible bladders in the propellant tanks. Helium routed into the void between the bladder and the tank wall squeezes the bladder to positively expel the propellant under zero-gravity conditions. The propellants flow through normally open main shutoff valves into separate fuel and oxidizer manifolds that lead to the thrusters. A switch on the control panel enables the astronauts to simultaneously close a pair of fuel and oxidizer main shutoff valves, thereby isolating a system's propellant tanks from its thrusters, if the propellants of that system are depleted or if the system malfunctions. After shutting off one system, the astronauts can restore operation of all 16 thrusters by opening the cross-feed valves between the system A and B manifolds.

During ascent engine firing, the astronauts may open the normally closed ascent propulsion section/RCS interconnect lines if the LM is accelerating in the +X-axis (upward) direction; closing the interconnect lines shortly before ascent engine shutdown ensures that no ascent helium enters the RCS propellant lines. Control panel switches open the interconnect valves in fuel-oxidizer pairs, for an individual RCS system, or for both systems simultaneously.

Transducers in the propellant tanks sense helium pressure and fuel temperature. Due to the proximity of the fuel tank to the oxidizer tank, the fuel temperature is representative of propellant temperature. Quantity indicators for system A and B display the summed quantities of fuel and oxidizer remaining in the tanks.

THRUSTER SECTION

Each of the four RCS clusters consists of a frame, four thrusters, eight heating elements, and associated sensors and plumbing. The clusters are diametrically opposed, evenly distributed around the ascent stage. The frame is an aluminum-alloy casting, shaped like a hollow cylinder, to which the four thrusters are attached; the entire cluster assembly is connected to the ascent stage by hollow struts. The vertical-firing thrusters are at the top and bottom of the cluster frame, the horizontal-firing thrusters are at each side. Each cluster is enclosed in a thermal shield; part of the four thruster combustion chambers and the extension nozzles protrude from the shield. The thermal shields aid in maintaining a temperature-controlled environment for the propellant lines from the ascent stage to the thrusters, minimize heat loss, and reflect radiated engine heat and solar heat.

The RCS thrusters are radiation-cooled, pressure-fed, bipropellant rocket engines that operate in a pulse mode to generate short thrust impulses for fine attitude corrections (navigation alignment maneuvers) or in a steady-state mode to produce continuous thrust for major attitude or translation changes. In the pulse mode, the thrusters are fired intermittently in bursts of less than 1 second duration — the minimum pulse may be as short as 14 milliseconds — however, the thrust level does not build up to the full 100 pounds that each thruster can produce. In the steady-state mode, the thrusters are fired continuously (longer than 1 second) to produce a stabilized 100 pounds of thrust until the shutoff command is received.

Two electric heaters, which encircle the thruster injector, control propellant temperature by conducting heat to the combustion chamber and the propellant solenoid valves. The heaters maintain the cluster at approximately +140° F, ensuring that the combustion chambers are properly preheated for instantaneous thruster starts. The astronauts can determine, by use of a temperature indicator and a related selector switch, if a cluster temperature is below the minimum operational temperature of 119° F and take corrective action to restore the cluster temperature.

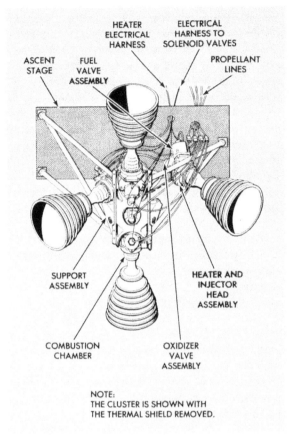

NOTE:
THE CLUSTER IS SHOWN WITH
THE THERMAL SHIELD REMOVED.

R-93

Thrust Chamber Cluster

Propellants are prevented from entering the thrusters by dual-coil, solenoid-operated shutoff valves at the fuel and oxidizer inlet ports. These valves are normally closed; they open when an automatic or a manual command energizes the primary or secondary coil, respectively. Seven milliseconds after receiving the thruster-on command, the valves are fully opened and the pressurized propellants flow through the injector into the combustion chamber where ignition occurs. By design, the fuel valve opens 2 milliseconds before the oxidizer valve, to provide proper ignition characteristics. Orifices at the valve inlets meter the propellant flow so that an oxidizer to fuel mixture ratio of 2 to 1 is obtained at the injector.

As the propellants mix and burn, the hot combustion gases increase the chamber pressure, accelerating the gas particles through the chamber exit. The gases are expanded through the divergent section of the nozzle at supersonic velocity, eventually building up to reach a reactive force of 100 pounds of thrust in the vacuum of space. The gas temperature within the combustion chamber stabilizes at approximately 5,200° F. The temperature at the nonablative chamber wall is maintained at a nominal 2,200° F by a combined method of film cooling (a fuel stream sprayed against the wall) and radiation cooling (dissipation of heat from the wall surface into space).

When the thruster-off command is received, the coils in the propellant valves deenergize, and spring pressure closes the valves. Propellant trapped in the injector is ejected and burned for a short time, while thrust decays to zero pounds.

When a thruster-on signal commands a very short duration pulse, engine thrust may be just beginning to rise when the pulse is ended and the propellant valves close. Under these conditions, the thrusters do not develop the full-capacity thrust of 100 pounds.

A failure-detection system informs the astronauts should a thruster fail on (fires without an on command) or off (does not fire despite an on command). Either type of failure produces the same indication: a warning light goes on and the talkback related to the failed thruster pair changes from the normal gray to a red display. The astronauts then disable the malfunctioning thruster pair by operating appropriate LGC thruster pair command switch and pulling associated circuit breaker (if thruster fail on condition exists). To offset the effects of a thruster-on failure, opposing thrusters will automatically receive fire commands and keep firing until the failed-on thruster has been disabled. A thruster-off condition is detected by a pressure switch, which senses combustion chamber pressure. When a fire command is received, the solenoid valves of the thruster open, resulting in ignition and subsequently in pressure buildup in the combustion chamber. When the pressure reaches 10.5 psia, the switch closes, indicating that proper firing is in process. When a very short duration fire command is received (a pulse of less than 80 milliseconds), the combustion chamber pressure may not build up enough for a proper firing. Short pulse skipping does not result in a failure indication, unless six consecutive pulses to the same thruster have not produced a response. In this case, the warning light and the talkback inform the astronauts that they have a nonfiring thruster, which must be isolated.

EQUIPMENT

EXPLOSIVE VALVES

The explosive valves are single-cartridge-actuated, normally closed valves. The cartridge is fired by applying power to the initiator bridgewire. The resultant heat fires the initiator, generating gases in the valve explosion chamber at an extremely high rate. The gases drive the valve piston into the housing, aligning the piston port permanently with the helium pressurization line.

PROPELLANT QUANTITY MEASURING DEVICE

The propellant quantity measuring device, consisting of a helium pressure/temperature probe and an analog computer for each system, measures the total quantity of propellants (sum of fuel and oxidizer) in the fuel and oxidizer tanks. The output voltage of the analog computer is fed to an indicator and is displayed to the astronauts on two scales (one for each RCS system) as percentage of propellant remaining in the tanks.

The propellant quantity measuring device uses a probe to sense the pressure/temperature ratio of the gas in the helium tank. This ratio, directly proportional to the mass of the gas, is fed to an analog computer that subtracts the mass in the helium tank from the total mass in the system, thereby deriving the helium mass in the propellant tanks. Finally, propellant tank ullage volume is subtracted from total tank volume to obtain the quantity of propellant remaining. Before firing the helium isolation explosive valves, the quantity displayed exceeds 100%, so that, after the valves are opened and the gas in the helium tank becomes less dense, the indicated quantity will be 100%.

PROPELLANT STORAGE TANKS

The four propellant tanks, one fuel and one oxidizer tank for each system, are cylindrical with hemispherical ends; they are made of titanium alloy. In each tank, the propellant is stored in a Teflon bladder, which is chemically inert and resistant to the corrosive action of the propellants. The bladder is supported by a standpipe running lengthwise in the tank. The propellant is fed into the tank from a fill point accessible from the exterior of the LM. A bleed line that extends up through the standpipe draws off gases trapped in the bladder. Helium flows between the bladder and the tank wall and acts upon the bladder to provide positive propellant expulsion.

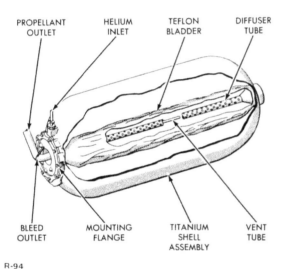

R-94

Propellant Storage Tank

THRUST CHAMBER ASSEMBLIES

The efficiency of a rocket engine is expressed in terms of specific impulse, which is the impulse-producing capacity per unit weight of propellant. The nominal specific impulse of the RCS thrusters at steady-state firing is 281 seconds. The thrusters have a favorable high-thrust to minimum-impulse ratio, meaning that they produce a comparatively high thrust for their size, as well as a very low thrust impulse. In addition, the thrusters have a fast response time. Response time is the elapsed time between a thruster-on command and stable firing at rated thrust, and between a thruster-off command and thrust decay to an insignificant value. Finally, the thrusters have a long cycle life, denoting that the thrusters can be restarted many times.

Each thruster consists of a fuel valve, an oxidizer valve, an injector head assembly, a combustion chamber, an extension nozzle, and thruster instrumentation.

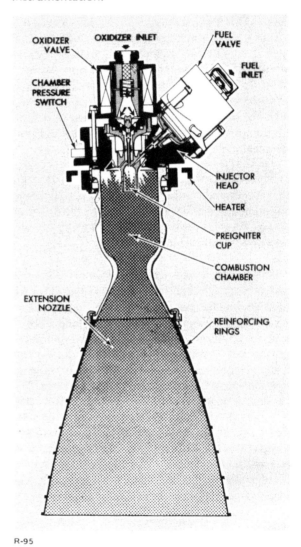

R-95

Reaction Control Thruster

The fuel and oxidizer valves are normally closed, two-coil, solenoid valves that control propellant flow to the injector. Each valve has an inlet filter, an inlet orifice, a spool assembly, a spring, an armature, and a valve seat. The primary and the secondary coils are wound on a magnetic core in the spool assembly. These coils receive the thruster on and off commands. The fuel and oxidizer valves are identical except for the inlet orifice, the valve seat, and the spool assembly. Because the ratio of oxidizer to fuel at the combustion chamber must be approximately 2 to 1, the diameters of the inlet orifices and the valve seat exits differ in the two valves. The spool assembly in the fuel valve produces a faster armature response to open the fuel valve 2 milliseconds before the oxidizer valve. Permitting fuel to enter the combustion chamber first reduces the possibility of ignition delay, which could cause temporary overpressurization (spiking) in the combustion chamber. Spiking is also held to a minimum by preheating and prepressurizing the combustion chamber.

When the thruster-off command is given, the coils deenergize, releasing the armature poppets. Spring and propellant pressure return the armature poppet of each valve to its seat, shutting off propellant flow into the injector.

The injector head assembly supports the fuel and oxidizer valves and the mounting flange for the combustion chamber. The propellant impingement and chamber cooling arrangement in the injector consists of four concentric orifice rings and a preigniter cup. Initial combustion occurs in the preigniter cup (a precombustion chamber) where a single fuel spray and oxidizer stream impinge. This provides a smoother start transient because it raises the main combustion chamber pressure for satisfactory ignition. The main fuel flow is routed through holes in a tube to a chamber that channels the fuel to an annulus. The annulus routes fuel to three concentric fuel rings. The outermost ring sprays fuel onto the combustion chamber wall, where it forms a boundary layer for cooling. The middle ring has eight orifices that spray fuel onto the outer wall of the preigniter cup to cool the cup. Eight primary orifices of the middle ring eject fuel to mix with the oxidizer. The main oxidizer flow is routed through holes in the oxidizer preigniter tube, to a chamber that supplies the eight primary oxidizer orifices of the innermost ring. The primary oxidizer and fuel orifices are arranged in doublets, at angles to each other, so that the emerging propellant streams impinge. Due to the hydraulic delay built into the injector, ignition at these eight doublets occurs approximately 4 milliseconds later than ignition inside the preigniter cup.

The combustion chamber is made of machined molybdenum, coated with silicon to prevent oxidation of the base metal. The chamber is cooled by radiation and by a film of fuel vapor. The extension nozzle is fabricated from L605 cobalt base alloy; eight stiffening rings are machined around its outer surface to maintain nozzle shape at high temperatures. The combustion chamber and extension nozzle are joined together by a large coupling nut and lockring.

HEATERS

Two redundant, independently operating heating systems are used simultaneously to heat the RCS clusters. Two electric heaters, one from each system, encircle the injector area of each thruster. The heaters normally operate in an automatic mode; redundant thermal switches (two connected in parallel for each thruster) sense injector temperature and turn the heaters on and off to maintain the temperature close to $+140^\circ$ F. The heaters of the primary heating system are powered directly from their circuit breakers. Power to the redundant system is routed through switches that permit the astronauts to operate this system for each cluster individually, either under automatic thermal switch control or with heaters continuously on, or off.

ELECTRICAL POWER

QUICK REFERENCE DATA

A-C Section
  Inverter input voltage — 24 to 32 volts dc
  Inverter output (with internal sync) — 115±1.2 volts rms, 400 Hz, single phase
  Normal load range — 0 to 350 volt-amperes (at power factors 0.65 lagging to 0.80 leading)
  Maximum overload at constant voltage output — 525 volt-amperes for 10 minutes

D-C Section
  Steady-state bus voltage limits — 26.5 to 32.5 volts dc
  Nominal supply bus voltage — 28 volts dc
  Transient voltages — 50 volts above or below nominal supply voltage

Descent battery
  Number of batteries — 5
  Capacity (each battery) — 415 ampere-hours
  Nominal voltage — 30.0 volts dc
  Minimum voltage — 28.0 volts dc
  Maximum voltage — 32.5 volts dc
  Weight — 135 pounds
  Construction — Silver-zinc plates, 20 cells
  Electrolyte — Potassium hydroxide

Ascent battery
  Number of batteries — 2
  Capacity — 296 ampere-hours
  Nominal voltage — 30.0 volts dc
  Minimum voltage — 27.5 volts dc
  Maximum voltage — 32.5 volts dc
  Weight — 125 pounds
  Construction — Silver-zinc plates, 20 cells
  Electrolyte — Potassium hydroxide

The Electrical Power Subsystem (EPS) is the principal source of electrical power necessary for the operation of the LM. The electrical power is supplied by seven silver-zinc batteries: five in the descent stage and two in the ascent stage. The batteries provide dc for the EPS d-c section; two solid-state inverters supply the a-c section. Both sections supply operating power to respective electrical buses, which supply all LM subsystems through circuit breakers. Other batteries supply power to trigger explosive devices, to operate the portable life support system, and to operate scientific equipment.

The descent stage batteries power the LM from T-30 minutes until the docked phases of the mission, at which time the LM receives electrical power from the CSM. After separation from the CSM, during the powered descent phase of the mission, the descent stage batteries are paralleled with the ascent stage batteries. Paralleling the batteries ensures the minimum required voltage for all possible LM operations. Before lift-off from the lunar surface, ascent stage battery power is introduced, descent battery power is terminated, and descent battery feeder lines are deadfaced and severed. Ascent stage battery power is then used

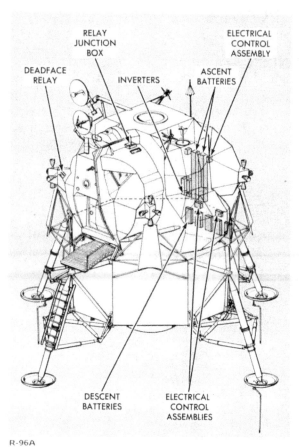

RELAY JUNCTION BOX

ELECTRICAL CONTROL ASSEMBLY

DEADFACE RELAY

ASCENT BATTERIES

INVERTERS

DESCENT BATTERIES

ELECTRICAL CONTROL ASSEMBLIES

R-96A

Major Electrical Power Equipment Location

until after final docking and astronaut transfer to the CM. The batteries are controlled and protected by electrical control assemblies, a relay junction box, and a deadface relay box, in conjunction with the control and display panel.

In addition to being the primary source of electrical power for the LM during the mission, the EPS is the distribution point for externally generated power during prelaunch and docked operations. Prelaunch d-c and a-c power is initially supplied from external ground power supplies until approximately T-7 hours. At this time, the vehicle ground power supply unit is removed and d-c power from the launch umbilical tower is connected. From launch until LM-CSM transition and docking, the EPS distributes internally generated d-c power. After docking, LM power is shut down and the CSM supplies d-c power to the LM. Before LM-CSM separation, all LM internally supplied electrical power is restored.

FUNCTIONAL DESCRIPTION

The outputs of the five descent stage batteries and two ascent stage batteries are applied to four electrical control assemblies. The two descent stage electrical control assemblies provide an independent control circuit for each descent battery. The two ascent stage electrical control assemblies provide four independent battery control circuits, two control circuits for each ascent battery. The electrical control assembly monitors reverse-current, overcurrent, and overtemperature within each battery. Each battery control circuit can detect a bus or feeder short. If an overcurrent condition occurs in a descent or ascent battery, the control circuit operates a main feed contactor associated with the malfunctioning battery to remove the battery from the distribution system.

Ascent and descent battery main power feeders are routed through circuit breakers to the d-c buses. From these buses, power is distributed through circuit breakers to all LM subsystems. The two inverters, which make up the a-c section power source, are connected to either of two a-c buses. Either inverter, when selected, can supply the LM a-c requirements.

Throughout the mission, the astronauts monitor the primary a-c and d-c voltage levels, d-c current levels, and the status of all main power feeders. The electrical power control and indicator panel in the cabin has talkbacks that indicate main power feeder status, indicators that display battery and bus voltages and currents, and component caution lights. The component caution lights are used to detect low bus voltages, out-of-limit, a-c bus frequencies, and battery malfunctions. Backup a-c and d-c power permits the astronauts to disconnect, substitute, or reconnect batteries, feeder lines, buses, or inverters to assure a continuous electrical supply.

EQUIPMENT

DESCENT STAGE BATTERIES

The five descent stage batteries are identical. Each battery is composed of silver-zinc plates, with

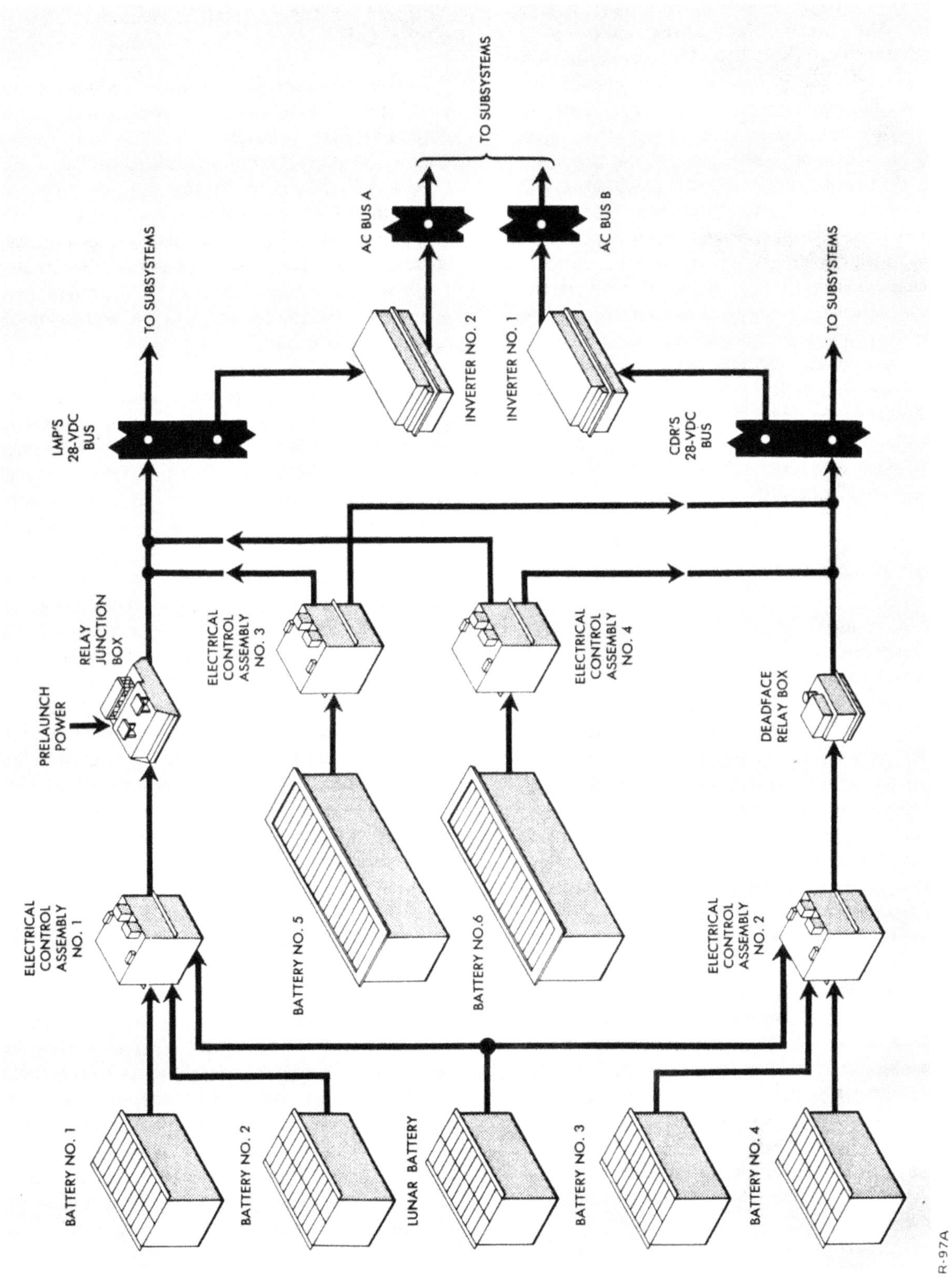

Diagram of Electrical Power Subsystem

R-97A

GRUMMAN

a potassium hydroxide electrolyte. Each battery has 20 cells, weighs 135 pounds, and has a 415-ampere-hour capacity (approximately 25 amperes at 28 volts dc for 16 hours, at +80° F). Normally, the descent stage batteries are paralleled so that they discharge equally. The batteries can operate in a vacuum while cooled by an Environmental Control Subsystem (ECS) cold rail assembly to which the battery heat sink surface is mounted. Five thermal sensors monitor cell temperature limits (+145° ±5° F) within each battery; they cause a caution light to go on to alert the astronaut to a battery over-temperature condition. The batteries initially have high-voltage characteristics; a low-voltage tap is provided (at the 17th cell) for use from T-30 minutes through transposition and docking. The high-voltage tap is used for all other normal LM operations. If one descent stage battery fails, the remaining descent stage batteries can provide sufficient power.

ASCENT STAGE BATTERIES

The two ascent stage batteries are identical. Each battery is composed of silver-zinc plates, with a potassium hydroxide electrolyte. Each battery weighs 125 pounds, and has a 296-ampere-hour capacity (50 amperes at 28 volts for 5.9 hours, at +80° F). To provide independent battery systems, the batteries are normally not paralleled during the ascent phase of the mission. The batteries can operate in a vacuum while cooled by ECS cold rails to which the battery heat sink surface is mounted. The nominal operating temperature of the batteries is approximately +80° F. Battery temperature in excess of +145° ±5° F closes a thermal sensor, causing a caution light to go on. The astronaut then takes corrective action to disconnect the faulty battery. The batteries ordinarily supply the d-c power requirements, from normal staging to final docking of the ascent stage with the orbiting CSM or during any malfunction that requires separation of the ascent and descent stages. If one ascent stage battery fails, the remaining battery provides sufficient power to accomplish safe rendezvous and docking with the CSM during any part of the mission.

DESCENT STAGE ELECTRICAL CONTROL ASSEMBLIES

The two descent stage electrical control assemblies control and protect the descent stage batteries. Each assembly has a set of control circuits for each battery accommodated. A failure in one set of battery control circuits does not affect the other set. The protective circuits of the assembly automatically disconnect a descent stage battery if an overcurrent condition occurs and cause a caution light to go on if a battery overcurrent, reverse-current, or overtemperature condition is detected.

The major elements of each assembly are high- and low-voltage main feed contactors, current monitors, overcurrent relays, reverse-current relays, and power supplies. An auxiliary relay supplies system logic contact closures to other control assemblies in the LM power distribution system.

The reverse-current relay causes a caution light to go on when current flow in the direction opposite to normal current flow exceeds 10 amperes for at least 4 seconds. Unlike the overcurrent relay, the reverse-current relay does not open the related main feed contactor and is self-resetting when the current monitor ceases to detect a reverse-current condition. During reverse-current conditions, the related contactor must be manually switched open. The control assembly power supplies provide ac for current-monitor excitation and regulated dc for the other circuits.

ASCENT STAGE ELECTRICAL CONTROL ASSEMBLIES

The two ascent stage electrical control assemblies individually control and protect the two ascent stage batteries in nearly the same manner as the descent stage control assemblies. Each assembly contains electrical power feed contactors, an overcurrent relay, a reverse-current relay, and a current monitor. Each ascent stage battery can be connected to its normal or backup main feeder line via the normal or the backup main feed contactor in its respective assembly. Both batteries are thereby connected to the primary d-c power buses. The normal feeder line has overcurrent protection; the backup feeder line does not.

RELAY JUNCTION BOX

The relay junction box provides the following:

Control logic and junction points for connecting external prelaunch power (via the launch umbilical tower) to the LM Pilot's d-c bus

Control and power junction points for connecting descent stage and ascent stage electrical control assemblies to the LM Pilot's d-c bus

Deadfacing (electrical isolation) of half of the power feeders between the descent and ascent stages.

The relay junction box controls the low-voltage contactors of batteries 1 and 4 (on and off) from the launch umbilical tower and CSM, and all low- and high-voltage descent power contactors (off) on receipt of an abort stage command. The junction box includes abort logic relays, which, when energized by an abort stage command, close the ascent stage battery main feed contactors and open the deadface main feed contactors and deadface relays. The deadface relay is manually opened and closed or automatically opened when the abort logic relays close. The deadface relay in the junction box deadfaces half of the main power feeders between the descent and ascent stages; the other half of the power feeders is deadfaced by the deadface relay in the deadface relay box. The ascent stage then provides primary d-c power to the LM.

DEADFACE RELAY BOX

The deadface relay box deadfaces those power feeders that are not controlled by the relay junction box, in the same manner as the relay junction box. Two individual deadfacing facilities (28 volts for each circuit breaker panel) are provided.

INVERTERS

Two identical redundant, 400-Hz inverters individually supply the primary a-c power required

in the LM. Inverter output is derived from a 28-volt d-c input. The output of the inverter stage is controlled by 400-Hz pulse drives developed from a 6.4-kilopulse-per-second (kpps) oscillator, which is, in turn, synchronized by timing pulses from the Instrumentation Subsystem. An electronic tap changer sequentially selects the output of the tapped transformer in the inverter stage, converting the 400-Hz square wave to an approximate sine wave of the same frequency. A voltage regulator maintains the inverter output at 115 volts ac during normal load conditions by controlling the amplitude of a dc-to-dc converter output. The voltage regulator also compensates for variations in the d-c input and a-c output load. When the voltage at a bus is less than 112 volts ac, or the frequency is less than 398 Hz or more than 402 Hz, a caution light goes on. The light goes off when the malfunction is remedied.

CIRCUIT BREAKER AND EPS CONTROL PANELS

All primary a-c and d-c power feed circuits are protected by circuit breakers on the Commander's and LM Pilot's buses. The two d-c buses are electrically connected by the main power feeder network. Functionally redundant LM equipment is placed on both d-c buses (one on each bus), so that each bus can individually perform a mission abort.

SENSOR POWER FUSE ASSEMBLIES

Two sensor power fuse assemblies, in the aft equipment bay, provide a secondary d-c bus system that supplies excitation to transducers in other subsystems that develop display and telemetry data. During prelaunch procedures, primary power is supplied to the assemblies from the Commander's 28-volt d-c bus. Before launch, power from the launch umbilical tower is disconnected, and power is subsequently available to the sensor power fuse assemblies from the LM Pilot's 28-volt d-c bus. Each assembly comprises a positive d-c bus, negative return bus, and 40 fuses. All sensor return lines are routed to a common ground bus.

COMMUNICATIONS

QUICK REFERENCE DATA

RF electronic equipment

    S-band transceiver assembly

        Frequency

            Transmit   2282.5 mHz (downlink)

            Receive   2101.8 mHz (uplink)

        Output power   0.75 watts (minimum)

        Input power requirement   36 watts

        Application   LM-MSFN communications

    S-band power amplifiers

        Frequency   2282.5 mHz

        Output power

            Primary amplifier   18.6 watts (minimum)

            Secondary amplifier   14.8 watts (minimum)

        Input power requirement   72 watts

        Application   Amplify S-band transmitter output

    VHF transceiver assembly

        Frequency

            Channel A   296.8 mHz

            Channel B   259.7 mHz

        Output power   5.0 watts

        Input power requirement

            VHF A transmitter   30 watts

            VHF A receiver   1.2 watts

            VHF B transmitter   31.7 watts

            VHF B receiver   1.2 watts

        Application   LM-CSM and LM-EVA communications

Signal-processing equipment

    Signal processor assembly

        Input power requirement   12.7 watts

        Application   Switching center for most signals in the Communications Subsystem

    Digital uplink assembly

        Input power requirements 12.5 watts

        Application   Processes MSFN signal to update LM guidance computer and provides MSFN voice backup

    Ranging tone transfer assembly

        Input power requirements 5.0 watts

        Application   Provides CSM-LM ranging in conjunction with VHF transceiver assembly

Microphones   Noise-cancelling, dynamic

Headsets   Dual muff (suits)

                               Ear plug (lightweight)

GRUMMAN

Signal-processing equipment (cont)

Television camera
    Bandwidth   10 Hz to 500 kHz
    Scan   10 fps, 320 lines
                                               5/8 fps, 1,280 lines
    Input power requirement   7.5 watts

Antenna equipment
    S-band steerable antenna
        Operating frequency
            Transmit   2282.5 mHz
            Receive   2101.8 mHz
        Type   Cross-sleeved dipoles over ground plane with parabolic reflector

        Slew movement
            Azimuth   174°
            Elevation   330°
    S-band in-flight antennas
        Operating frequency
            Transmit   2282.5 mHz
            Receive   2101.8 mHz
        Type   Omnidirectional, right-hand circularly polarized

    VHF In-flight antennas
        Operating frequency   259.7 and 296.8 mHz
        Type   Omnidirectional, right-hand circularly polarized
    VHF EVA antenna
        Operating frequency   259.7 to 296.8 mHz
        Type   Omnidirectional, conical, 8-inch monopole with 12-inch radials

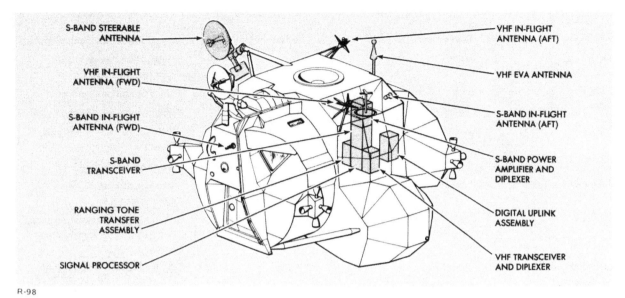

R-98

Major Communications Equipment Locations

The Communications Subsystem (CS) provides in-flight and lunar surface communications links between the LM and CSM, the LM and MSFN, and the LM and the extravehicular astronaut (EVA). When both astronauts are outside the LM, the LM relays communications between the astronauts and MSFN. When the astronauts are in the Lunar Roving Vehicle (LRV), the Lunar Communications Relay Unit (LCRU), mounted on the LRV, is the communications relay. The CS consists of S-band and VHF equipment.

IN-FLIGHT COMMUNICATIONS

In flight, when the LM is separated from the CSM and is on the earth side of the moon, the CS provides S-band communications with MSFN and VHF communications with the CSM. When the LM and the CSM are on the far side of the moon, VHF is used for communications between them.

EARTH SIDE (LM-MSFN)

In-flight S-band communications between the LM and MSFN include voice, digital uplink signals, and ranging code signals from MSFN. The LM S-band equipment transmits voice, acts as transponder to the ranging code signals, transmits biomedical and systems telemetry data, and provides a voice backup capability and an emergency key capability.

S-band voice is the primary means of communication between MSFN and the LM. Backup voice communication from MSFN is possible, using the digital uplink assembly, but this unit is normally used by the MSFN to update the LM guidance computer. In response to ranging code signals sent to the LM, the S-band equipment supplies MSFN with a return ranging code signal that enables MSFN to track, and determine the range of the LM. The LM transmits biomedical data pertinent to astronaut heartbeat so that MSFN can monitor and record the physical condition of the astronauts. The LM also transmits systems telemetry data for MSFN evaluation; voice, using redundant S-band equipment; and, in case there is no LM voice capability, provides an emergency key signal so that the astronauts can transmit Morse code to MSFN.

EARTH SIDE (LM-CSM)

In-flight VHF communications between the LM and CSM include voice, backup voice, and tracking and ranging signals. Normal LM-CSM voice communications use VHF channels A and B duplex. Backup voice communication is accomplished with VHF channel B simplex or channel A simplex VHF ranging, initiated by the CSM, uses VHF channels A and B duplex.

Link	Mode	Band	Purpose
MSFN-LM-MSFN	Pseudorandom noise (PRN)	S-band	Ranging and tracking by MSFN
LM-MSFN	Voice	S-band	In-flight and lunar surface communications
LM-CSM	Voice	VHF duplex	In-flight communications
CSM-LM-MSFN	Voice	VHF and S-band	Conference (with LM as relay)
LM-CSM	Low-bit-rate telemetry	VHF (one way)	CSM records and retransmits to earth
CSM-LM-CSM	Ranging	VHF duplex	Ranging by CSM
MSFN-LM	Voice	S-band	In-flight and lunar surface communications
MSFN-LM	Uplink data or uplink voice backup	S-band	Update LM guidance computer or voice backup for in-flight communications
LM-MSFN	Television	S-band	Provides lunar color television
LM-MSFN	Biomed-PCM telemetry	S-band	Transmission of biomedical and vehicle status data
LM-MSFN-CSM	Voice	S-band	Conference (with earth as relay)
EVA-EVA-LM	Voice and data; voice	VHF duplex	EVA direct communication
EVA-LM-MSFN	Voice and data	VHF, S-band	Conference (with LM as relay)
CSM-MSFN-LM-EVA	Voice and data	S-band, VHF	Conference (via MSFN-LM relay)

Communication Links

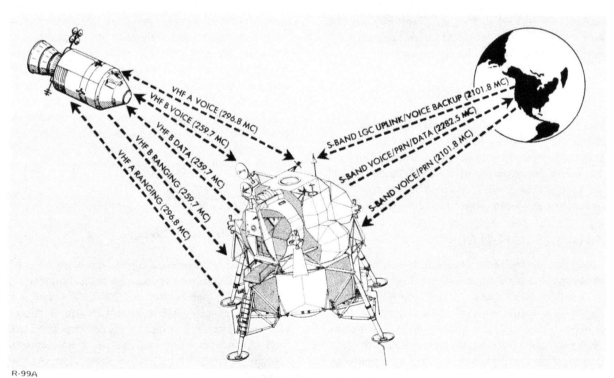

In-Flight Communications

R-99A

GRUMMAN

FAR SIDE (LM-CSM)

When the LM and CSM are behind the moon, contact with MSFN is not possible. VHF channels A and B are used for duplex LM-CSM voice communications. VHF channel B is used as a one-way data link to transmit system telemetry signals from the LM, to be recorded and stored by the CSM. When the CSM establishes S-band contact with MSFN, the stored data are transmitted by the CSM at 32 times the recording speed.

LUNAR SURFACE COMMUNICATIONS

When the LM is on the lunar surface, the CS provides S-band communications with MSFN and VHF communications with the EVA. The LM relays VHF signals to MSFN, using the S-band.

Communications with the CSM may be accomplished by using MSFN as a relay. LM-MSFN S-band capabilities are the same as in-flight capabilities, except that, in addition, TV may be transmitted from the lunar surface in an FM mode.

Information	Frequency or Rate	Subcarrier Modulation	Subcarrier Frequency	RF Carrier Modulation
UPLINK: 2101.8 mHz				
Voice	300 to 3000 Hz	FM	30 kHz	PM
Voice backup	300 to 3000 Hz	FM	70 kHz	PM
PRN ranging code	990.6 kilobits/sec			PM
Uplink data	1.0 kilobits/sec	FM	70 kHz	PM
DOWNLINK: 2282.5 mHz				
Voice	300 to 3,000 Hz	FM	1.25 mHz	PM or FM
Biomed	14.5-kHz subcarrier	FM	1.25 mHz	PM or FM
Extravehicular mobility unit	3.9-, 5.4-, 7.35- and 10.5-kHz subcarriers	FM	1.25 mHz	PM or FM
Voice	300 to 3000 Hz	None	None	Direct PM baseband modulation
Extravehicular mobility unit	3.9-, 5.4-, 7.35-, and 10.5-kHz subcarriers	None	None	Direct PM baseband modulation
Voice backup	300 to 3000 Hz	None	None	Direct PM baseband modulation
PRN ranging code (turnaround)	990.6 kilobits/sec			PM
Emergency keying	Morse code	AM	512 kHz	PM
Pulse-code-modulation nonreturn-to-zero data	High bit rate: 51.2 kilobits/sec or Low bit rate: 1.6 kilobits/sec	Phase shift keying (PSK)	1.024 mHz	PM or FM
TV	10 to 500 Hz			FM baseband modulation

S-Band Communications Capabilities

Lunar Surface Communications

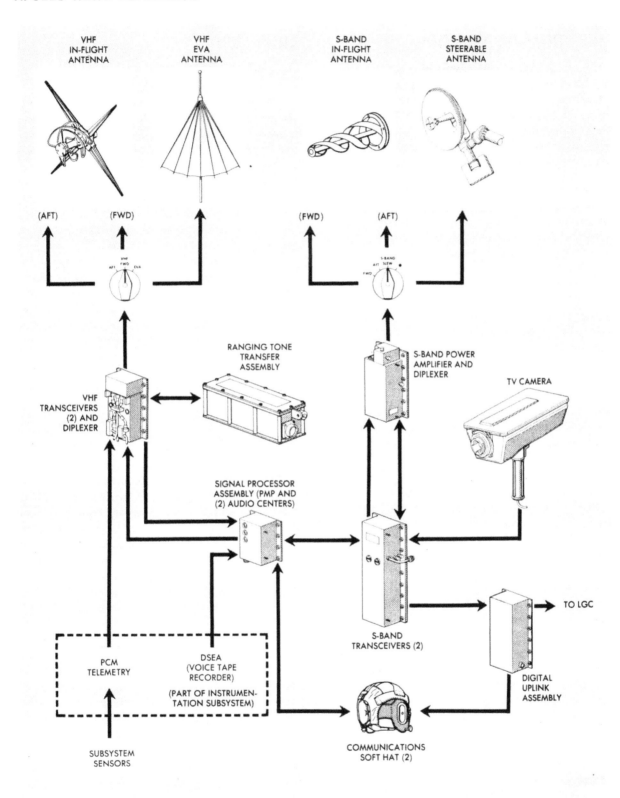

VHF
IN-FLIGHT
ANTENNA

VHF
EVA
ANTENNA

S-BAND
IN-FLIGHT
ANTENNA

S-BAND
STEERABLE
ANTENNA

(AFT)

(FWD)

(FWD)

(AFT)

RANGING TONE
TRANSFER
ASSEMBLY

S-BAND POWER
AMPLIFIER AND
DIPLEXER

TV CAMERA

VHF
TRANSCEIVERS
(2) AND
DIPLEXER

SIGNAL PROCESSOR
ASSEMBLY (PMP AND
(2) AUDIO CENTERS)

TO LGC

S-BAND
TRANSCEIVERS (2)

PCM
TELEMETRY

DSEA
(VOICE TAPE
RECORDER)

(PART OF INSTRUMEN-
TATION SUBSYSTEM)

DIGITAL
UPLINK
ASSEMBLY

SUBSYSTEM
SENSORS

COMMUNICATIONS
SOFT HAT (2)

R-101A

Diagram of Communications Subsystem

FUNCTIONAL DESCRIPTION

Each astronaut has his own audio center. The audio centers have audio amplifiers and switches that are used to route signals between the LM astronauts, and between the LM and MSFN or the CSM. The centers are redundant in that each one can be used by either astronaut, or both astronauts can use either audio center if necessary.

In a transmission mode, the output of the audio centers goes to the VHF transceivers, or to the premodulation processor or to the data storage electronics assembly in the Instrumentation Subsystem (IS). If an audio center output is routed to the VHF transmitter, the transmission is through the diplexer to the selected VHF antenna. If an audio center output is routed to the premodulation processor (PMP) and then to the S-band transceivers, the transmitter output is applied to the diplexer, or to the S-band power amplifier, depending on power output requirements. The output from the transmitter or the power amplifier goes through the diplexer to the selected S-band antenna. If an audio center output is routed to the data storage electronics assembly, the voice transmission is recorded.

The inputs to the S-band transceivers are from the premodulation processor or the television camera. The outputs from the premodulation processor (to be transmitted by S-band transmitters) are processed voice, and PCM, EMU and biomed data. For television transmission, the S-band power amplifier is used. In normal flight, and on the lunar surface, the steerable antenna is used. When the LRV is in use, transmission is through the S-band antenna mounted on it. The S-band omni antennas are used in any one of a number of backup modes.

External RF inputs to the S-band equipment are MSFN voice, either uplink data or an uplink backup voice signal, and ranging. Received MSFN voice is routed through the premodulation processor to the audio centers. Received uplink data signals are routed to the digital uplink assembly to be decoded and sent to the LM guidance computer. MSFN backup voice is routed to the digital uplink assembly where it is decoded and then sent to the Commander's microphone amplifier input.

EQUIPMENT

S-BAND TRANSCEIVER

The S-band transceiver assembly provides deep-space communications between the LM and MSFN. S-band communications consist of voice and pseudorandom noise ranging transmission from MSFN to the LM and voice, pseudorandom noise ranging turnaround, biomed, and subsystem data transmission from the LM to MSFN. The assembly consists of two identical phase-locked receivers, two phase modulators with driver and multiplier chains, and a frequency modulator. The receivers and phase modulators provide the ranging, voice, emergency-keying, and telemetry transmit-receive functions. The frequency modulator is primarily provided for video transmission, but accommodates pulse-code-modulation telemetry (subsystem data), biomed, and voice transmission. The frequency modulator provides limited backup for both phase modulators. The operating frequencies of the S-band equipment are 2282.5 mHz (transmit) and 2101.8 mHz (receive).

S-BAND POWER AMPLIFIER

The S-band power amplifier amplifies the S-band transmitter output when additional transmitted power is required. This assembly consists of two amplitrons, an input and an output isolator (ferrite circulators), and two power supplies, all mounted on a common chassis. The RF circuit is a series interconnection of the isolators and amplitrons. The amplitrons (which are characteristic of saturated, rather than linear, amplifiers) have broad bandwidth, high efficiency, high peak and average power output, but relatively low gain. The isolators protect both amplitrons and both S-band transmitter driver and multiplier chains. The isolators exhibit a minimum isolation of 20 db and a maximum insertion loss of 0.6 db. Each amplitron has its own power supply, One amplitron is designated primary; the other, secondary. Only one amplitron can be activated at a time. When neither amplitron is selected, a feedthrough path through the power amplifier exists with maximum insertion loss of 3.2 db (feedthrough mode).

VHF TRANSCEIVER

The VHF transceiver assembly provides voice communications between the LM and the CSM and, during blackout of transmission to MSFN, low-bit-rate telemetry transmission from the LM to the CSM, and ranging on the LM by the CSM. When the LM mission profile includes extra-vehicular activity, this equipment also provides EVA-LM voice communications, and reception of EVA biomed and suit data for transmission to MSFN over the S-band. The assembly consists of two solid-state superheterodyne receivers and two transmitters. One transmitter-receiver combination provides a 296.8-mHz channel (channel A); the other, a 259.7-mHz channel (channel B), for simplex or duplex voice communications. Channel B may also be used to transmit pulse-code-modulation data from the IS at the low bit rate and to receive biomed and suit data from the EVA during EVA-programmed missions.

SIGNAL PROCESSOR ASSEMBLY

The signal processor assembly is the common acquisition and distribution point for most CS received and transmitted data, except that low-bit-rate, split-phase data are directly coupled to VHF channel B and TV signals are directly coupled to the S-band transmitter. The signal processor assembly processes voice and biomed signals and provides the interface between the RF electronics, data storage electronics assembly, and pulse-code-modulation and timing electronics assembly of the IS. The signal processor assembly consists of an audio center for each astronaut and a premodulation processor. The signal processor assembly does not handle ranging and uplink data signals. The premodulation processor provides signal modulation, mixing, and switching in accordance with the selected mode of operation. It also permits the LM to be used as a relay station between the CSM and MSFN, and, for EVA-programmed missions, between the EVA and MSFN. The audio centers are identical. They provide individual selection, isolation, and amplification of audio signals received by the CS receivers and which are to be transmitted by the CS transmitters. Each audio center contains a microphone amplifier, headset amplifier, voice operated relay (VOX) circuit, diode switches, volume control circuits, and isolation pads. The VOX circuit controls the microphone amplifier by activating it only when required for voice transmission. Audio signals are routed to and from the VHF A, VHF B, and S-band equipments and the intercom bus via the audio centers. The intercom bus, common to both audio centers, provides hardline communications between the astronauts. Voice signals to be recorded by the data storage electronics assembly are taken from the intercom bus.

DIGITAL UPLINK ASSEMBLY

The digital uplink assembly decodes S-band uplink commands from MSFN and routes the data to the LM guidance computer. The digital uplink assembly provides a verification signal to the IS for transmission to MSFN, to indicate that the uplink messages have been received and properly decoded by the digital uplink assembly. The LM guidance computer also routes a no-go signal to the IS for transmission to MSFN whenever the computer receives an incorrect message. The uplink commands addressed to the LM parallel those inputs available to the LM guidance computer via the display and keyboard assembly. The digital uplink assembly also provides a voice backup capability if the received S-band audio circuits in the premodulation processor fail.

RANGING TONE TRANSFER ASSEMBLY

The ranging tone transfer assembly operates with VHF receiver B and VHF transmitter A to provide a transponder function for CSM-LM VHF ranging. The ranging tone transfer assembly receives VHF ranging tone inputs from VHF receiver B and produces ranging tone outputs to key VHF transmitter A.

The VHF ranging tone input consists of two acquisition tone signals and one track tone signal. Accurate ranging is accomplished when the track tone signal from the CSM is received and retransmitted from the LM.

S-BAND STEERABLE ANTENNA

The S-band steerable antenna is a 26-inch-diameter parabolic reflector with a point source feed that consists of a pair of cross-sleeved dipoles over a ground plane. The prime purpose of this antenna is to provide deep-space voice and telemetry communications and deep-space tracking and ranging. This antenna provides 174° azimuth coverage and 330° elevation coverage. The antenna can be operated manually or automatically. The manual mode is used for initial positioning of the antenna to orient it within ±12.5° (capture angle) of the line-of-sight signal received from the MSFN station. Once the antenna is positioned within the capture angle, it can operate in the automatic mode.

S-BAND IN-FLIGHT ANTENNAS

The two S-band in-flight antennas are omni-directional; one is forward and one is aft on the LM. The antennas are right-hand circularly polarized radiators that collectively cover 90% of the sphere at -3 db or better. They operate at 2282.5 mHz (transmit) and 2101.8 mHz (receive). These antennas are the primary S-band antennas for the LM when in flight.

VHF IN-FLIGHT ANTENNAS

The two VHF in-flight antennas are omni-directional, right-hand, circularly polarized antennas that operate at 259.7 and 296.8-mHz.

VHF EVA ANTENNA

The VHF EVA antenna is an omnidirectional conical antenna, which is used for LM-EVA communications when the LM is on the lunar surface. It is mounted on the LM and unstowed by an astronaut in the LM after landing.

INSTRUMENTATION

QUICK REFERENCE DATA

Signal-conditioning electronics assembly
- Height — 8.0 inches
- Width — 5.25 inches
- Length — 23.90 inches
- Weight
    - Assembly 1 — 35.44 pounds
    - Assembly 2 — 35.25 pounds
- Power requirements
    - Excitation — 28 volts dc
    - Consumption
        - Assembly 1 — 16.04 watts
        - Assembly 2 — 14.23 watts
- Temperature
    - Operating — +30° to +130° F
    - Nonoperating — -65° to +160° F

Pulse-code-modulation and timing
electronics assembly
- Height — 6.72 inches
- Width — 5.12 inches
- Length — 19.75 inches
- Weight — 23.0 pounds (approximate)
- Power requirements
    - Excitation — 28 volts dc
    - Consumption — 11 watts
- Operating temperature (ambient) — +30° to +130° F
- Number of analog channels — 277
- Normal bit rate (51.2 kilobits per second) — 200 channels externally programmed, 77 channels internally redundant
- Reduced bit rate (1.6 kilobits per second) — 113 channels externally programmed, 41 channels internally redundant

- Parallel digital signals
    - Number of channels — 75
    - Normal bit rate — 1, 10, 50, 100, or 200 samples per second
    - Reduced bit rate — 1 sample per second
- Serial digital signals
    - Number of channels — 2 channels, serial by bit
    - Normal bit rate — 50 samples per second
    - Reduced bit rate — None

Data storage electronics assembly
- Height — 2.05 inches
- Width — 4.0 inches
- Length — 6.22 inches
- Weight — 38 ounces
- Power supply input — 115 ± 2.5 volts rms, 400 Hz, single phase
- Magnetic heads — Two record/reproduce heads to provide four tracks
- Voice record amplifier
    - Input level — -3 to +7 dbm
    - Frequency response — ± 3 db from 300 Hz to 3 kHz

Tape
  Speed   0.6 inch per second
    Total recording time   10 hours (maximum)
    Length of tape between sensor strips   450 feet (minimum)
Transport
    Speed error   0.05 of input power frequency deviation
    Record time   Total of 10 hours
Caution and warning electronics assembly
  Height   7.0 inches
  Width   6.750 inches
  Depth   11.750 inches
  Weight   18.20 pounds
  Power requirements
    Excitation   28 volts dc
    Consumption   13 watts
  Temperature
    Operating   +35° to +135° F
    Nonoperating   −65° to +160° F

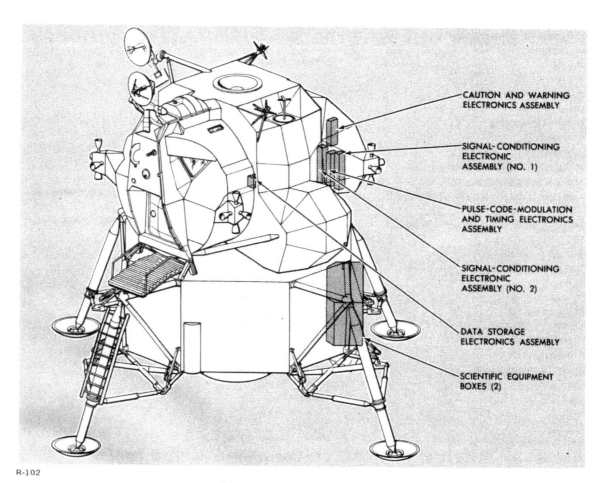

CAUTION AND WARNING
ELECTRONICS ASSEMBLY

SIGNAL-CONDITIONING
ELECTRONIC
ASSEMBLY (NO. 1)

PULSE-CODE-MODULATION
AND TIMING ELECTRONICS
ASSEMBLY

SIGNAL-CONDITIONING
ELECTRONIC
ASSEMBLY (NO. 2)

DATA STORAGE
ELECTRONICS ASSEMBLY

SCIENTIFIC EQUIPMENT
BOXES (2)

R-102

Major Instrumentation Equipment Location

The Instrumentation Subsystem (IS) monitors the LM subsystems, performs in-flight checkout, prepares LM status data for transmission, provides timing frequencies and correlated data for LM subsystems, stores voice and time-correlation data, performs lunar surface checkout, and provides scientific instrumentation for lunar experiments.

The IS monitors various parameters (status) of LM subsystems and structure and prepares the status data for telemetering via the Communications Subsystem (CS), to MSFN. In a high-bit-rate mode of operation, MSFN receives 51,200 bits of information from 279 subsystem sensors every second. This, along with Guidance, Navigation, and Control Subsystem data, enables mission controllers to participate in major decisions, assist in spacecraft management during complicated astronaut activity, and maintain a detail subsystem performance history.

Caution and warning lights and two master alarm lights alert the astronauts to out-of-tolerance conditions (malfunctions) that affect the mission or their safety. In addition a 3-kHz alarm tone is routed to the astronaut headsets to advise the astronauts that a malfunction exists. The tone is especially helpful in alerting the astronauts when they are preoccupied or asleep. The master alarm lights can be turned off by pushing either illuminated lens; this also stops the tone. When a warning light (red) goes on, it indicates a malfunction that affects the mission, but could affect astronaut safety if not corrected.

FUNCTIONAL DESCRIPTION

The IS consists of subsystem sensors, a signal-conditioning electronics assembly, a pulse-code-modulation and timing electronics assembly, a caution and warning electronics assembly, and a data storage electronics assembly.

The sensors continuously monitor the status of LM subsystems and provide outputs indicative of temperature, pressure, frequency, gas and liquid quantity, stage-separation distance, valve and switch positions, voltage, and current. These outputs are in analog and digital form; some are routed to the signal-conditioning electronics assembly for voltage-level conditioning. If conditioning is not required, the outputs are routed directly to the pulse-code-modulation and timing electronics assembly. The signal-conditioning electronics assembly conditions its sensor-derived inputs and routes high-level analog or digital data to the pulse-code-modulation and timing electronics assembly, caution and warning electronics assembly, and crew displays.

The pulse-code-modulation and timing electronics assembly converts the conditioned and unconditioned signals to several forms for telemetering. This assembly also provides subcarrier frequencies, time reference signals, and sync pulses.

The sensed subsystem data, routed in analog and digital form to the caution and warning electronics assembly, are constantly compared with internally generated references. When an out-of-tolerance condition is detected, this assembly provides a signal to light the appropriate warning or caution light and both master alarm lights and to provide the 3-kHz alarm tone to the headsets.

Basically, all caution and warning lights operate in the same manner. The following is a typical example. Signals are routed from Reaction Control Subsystem (RCS) helium tank pressure sensors to comparators in the caution and warning electronics assembly. If comparison indicates a low-pressure condition, solid-state electronic circuits are enabled, causing the RCS caution light to go on. The astronauts then monitor helium tank pressure on indicators to determine actual pressure levels.

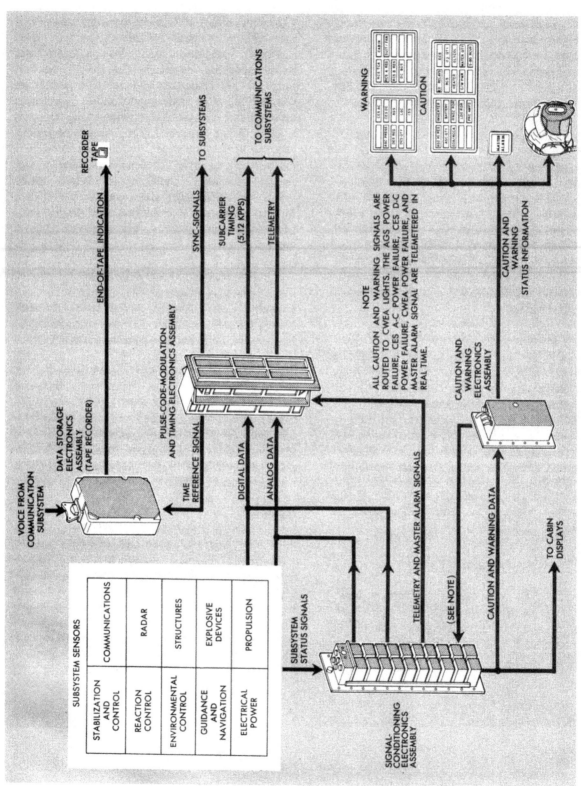

Diagram of Instrumentation Subsystem

R-103

Grumman

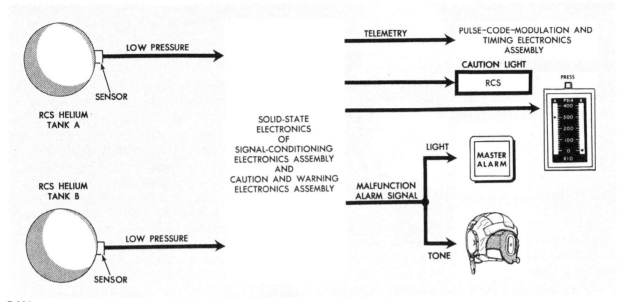

R-104

RCS Failure Detection

The data storage electronics assembly is a tape recorder that records voice and time-correlation data (mission elapsed time). The voice and data inputs are multiplexed and recorded. The recorder can be operated manually or semiautomatically. In the manual mode, an astronaut closes a push-to-talk switch on his attitude controller assembly or electrical umbilical and speaks into his microphone. In the semiautomatic mode, CS equipment senses voice inputs from within the cabin or from the communications receivers and activates the recorder. Voice signals from the CS intercom bus are also recorded, together with mission elapsed time.

EQUIPMENT

SUBSYSTEM SENSORS

The sensors fall into four general categories: mechanical, resistive, variable reluctance, and elec-trical. They are located throughout the various LM subsystems and structure and are used to change physical data into electrical signals.

SIGNAL CONDITIONING ELECTRONICS ASSEMBLY

This assembly consists of two electronic replace-able assemblies, each capable of housing up to 22 plug-in subassemblies of 11 different types (conver-ters, amplifiers, etc.). Each subassembly contains its own power supply, which is isolated from the other subassemblies. Loss of one subassembly, due to a power supply failure does not affect operation of the other subassemblies. The subassemblies perform one or more of the following seven functions: amplify d-c voltages, attenuate d-c voltages, convert ac to dc, convert frequency to dc, phase-modulate ac to dc, convert resistance varia-tions to d-c voltages, and isolate signals.

Pulse-Code-Modulation and Timing Electronics Assembly

This assembly comprises two sections: timing electronics and pulse code modulation. The timing electronics section develops timing signals for the pulse code modulation section, and the LM subsystems including the mission elapsed timer. The pulse-code-modulation section converts analog and digital signals to one of two formats, normal and reduced, for telemetering: 51,200 bits per second and 1600 bits per second.

Data Storage Electronics Assembly

This assembly is a single-speed, four-track, magnetic tape recorder that stores voice and time-correlation data. A maximum of 10 hours of recording time is provided (2.5 hours on each track) by driving the tape, at 0.6 inch per second, over the record head and, on completion of a pass, automatically switching to the next track and reversing tape direction. One tape (450 feet) is supplied in a magazine.

Caution and Warning Electronics Assembly

This assembly compares analog signals (between 0 and 5 volts dc), from the signal-conditioning electronics assembly, with preselected internally generated limits supplied by the caution and warning power supply as reference voltages. In addition to analog inputs, it receives discrete on-off and contact closure signals. All inputs are routed to detectors; the detected signals are routed through logic circuitry, enabling relay contacts that cause caution or warning lights to go on or causing talk-backs to change state. Simultaneously, the detected signal energizes a master relay driver, enabling relay contacts. These contacts route a signal to light the master alarm lights and trigger the 3-kHz tone to the headsets.

LIGHTING

QUICK REFERENCE DATA

Exterior tracking light
   Type — High-intensity, flashing
   Intensity — 9,000 beam candlepower (minimum)
   Visibility — Cone centered on LM +Z-axis, with semivertex angle of 30°
      Visual: 10 to 130 nautical miles
      CSM sextant: 30 to 400 nautical miles

Docking lights
   Type — Incandescent
   Intensity — Fixed
   Visibility — 1,000 feet

Interior
Control panels and pushbuttons	White
Circuit breakers	White
Numeric readouts	Green
Lunar contact lights	Blue
Caution annunciators	Yellow
Warning annunciators	Red
Master alarm pushbutton/lights	Red
Component caution lights	Yellow
Engine start and stop pushbutton/lights	Red
Computer status condition indicators	White
Self-luminous devices	Green
Talkbacks (two- and three-position)	White background
Displays	
Characters and indicia	White
Labels and multipliers	Green
Range markings	Green
Immediate-action or emergency controls	Yellow
Indicator power failure lights	Red
Floodlights	White

Exterior and interior lighting aids in the performance of crew visual tasks and lessens astronaut fatigue and interior-exterior glare effects. Exterior lighting is used for LM and CSM tracking and docking maneuvers. Interior lighting illuminates the cabin and the controls and displays on the Commander's and LM Pilot's panels.

FUNCTIONAL DESCRIPTION

LM lighting is provided by exterior and interior lights and lighting control equipment. The exterior lighting enables the astronauts to guide and orient the LM visually to the CSM visually to achieve successful tracking and docking. Interior lighting is divided into seven categories: incandescent annunciators, component caution lights, floodlights, computer condition lights, integral electroluminescent lighting, numeric electroluminescent lighting, and incandescently illuminated pushbuttons.

EXTERIOR LIGHTING

Exterior lighting includes five docking lights, and a high-intensity tracking light.

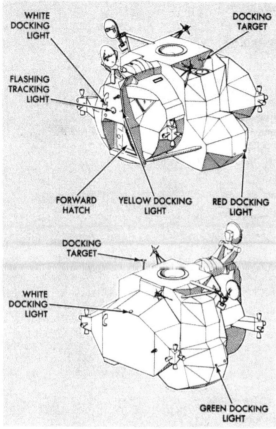

WHITE DOCKING LIGHT

DOCKING TARGET

FLASHING TRACKING LIGHT

FORWARD HATCH YELLOW DOCKING LIGHT RED DOCKING LIGHT

DOCKING TARGET

WHITE DOCKING LIGHT

GREEN DOCKING LIGHT

R-105

Ascent Stage Exterior Lighting

DOCKING LIGHTS

Five docking lights mounted on the exterior of the LM provide visual orientation and permit gross attitude determination relative to a line of sight through the CSM windows during rendezvous and docking. For transposition and docking, the docking lights are turned on by a switch located at spacecraft Lunar Module adapter attachment points. This switch is automatically closed upon deployment of the adapter panels. At completion of the docking maneuver, LM power is turned off and the docking lights go off. The lights are visible, and their color recognizable, at a maximum distance of 1,000 feet.

TRACKING LIGHT

The tracking light permits visual tracking of the LM by the CSM. A flash tube in the tracking light

electronics assembly causes the light, which has a 60° beam spread, to flash at a rate of 60 flashes per minute.

INTERIOR LIGHTING

Interior lighting consists of integral panel and display lighting, backup floodlighting, and electroluminescent lighting. Electroluminescence is light emitted from a crystalline phosphor (Z_NS) placed as a thin layer between two closely spaced electrodes of an electrical capacitor; one of the electrodes must be transparent. The light output varies with voltage. Advantageous characteristics are an "afterglow" of less than 1 second, low power consumption, and negligible heat dissipation.

INTEGRALLY LIGHTED COMPONENTS

There are three types of integrally lighted components: panel areas, displays, and caution and warning annunciators. The integrally lighted components use electroluminescent or incandescent devices that are controlled by on-off switches and potentiometer-type dimming controls. All panel placards are integrally lighted by white electroluminescent lamps with overlays. The displays have electroluminescent lamps within their enclosures. The numeric displays show green or white illuminated digits on a nonilluminated background; displays with pointers have a nonilluminated pointer travelling over an illuminated background. The brightness of the electroluminescent displays is varied with dimming controls which can be bypassed by a related override switch, so that full brightness will be maintained should a dimming control fail.

LUNAR CONTACT LIGHTS

Two Lunar Contact lights go on when one or more of the four lunar-surface sensing probes contact the lunar surface. A probe is mounted beneath each of the landing gear footpads.

FLOODLIGHTING

Floodlighting is used for general cabin illumination and as a secondary source of illumination for the control and display panels. Floodlighting is provided by the Commander's overhead and forward floodlights, the LM Pilot's overhead and forward floodlights, and recessed floodlights in the bottom of extending side panels. These floodlight fixtures provide an even distribution of light with minimum reflection. Every panel area has more than one lamp.

PORTABLE UTILITY LIGHTS

Two portable utility lights are used, when necessary, to supplement the cabin interior lighting. The lights, when removed from the flight data file container, connect to the overhead utility light panel. Switches provide one-step dimming for light-intensity control.

OPTICAL SIGHT RETICLE LIGHT

The crewman's optical alignment sight, used to sight the docking target on the CSM, has a reticle that is illuminated by a 28-volt d-c lamp.

ALIGNMENT OPTICAL TELESCOPE LIGHTS

A thumbwheel on the computer control and reticle dimmer assembly controls the brightness of the telescope reticle. The lamps edge-light the reticle with incandescent red light.

PORTABLE LIFE SUPPORT SYSTEM

The portable life support system provides an astronaut with a livable atmosphere inside his space suit during excursions on the lunar surface and in space. Worn on the back and connected to the suit's waist by umbilicals, it permits up to seven hours of extravehicular activity, depending on the astronauts metabolic rate.

The backpack supplies oxygen for breathing and suit ventilation, and refrigerated water and oxygen for body cooling. It pressurizes the suit to 3.9 psi and removes contaminants from the oxygen circulating through the suit. It also has a communication-telemetry set, controls to operate it, and devices to monitor its functions.

For the lunar mission, the LM has two of these life support packs. The LM carries enough supplies to refill each pack's oxygen tank and water reservoir, and replace its battery and two lithium hydroxide cannisters twice. This will allow a total of three extravehicular trips.

R-113A

Portable Life Support System

The life support pack, with its controls, weighs 104 pounds; it is 26 inches high, 20.5 inches wide, and 10.5 inches deep. It is powered by a 16.8-volt silver-zinc battery. A fiberglass cover protects the pack against micrometeroroids.

Five subsystems make up the portable life support system: primary oxygen supply, oxygen-ventilating circuit, water transport loop, feedwater loop, and space suit communication system. For emergency use, an oxygen purge system supplies an additional 30 to 90 minutes of oxygen according to selected flow rate. The OPS is mounted on the pack, but operates separately.

A thermal insulator made of fire-resistant Beta cloth and aluminized Kapton covers the pack and its shell to restrict heat leakage in or out, depending on the moon's temperature. A similar insulator covers the oxygen purge system.

A remote control unit (RCU), which is attached to the suit chest, has switches for the life support pack's water pump and oxygen fan, four-position communication selector switch, a radio volume control, an oxygen quantity gage, and warning indicators. The OPS actuator is attached to the RCU during an EVA.

FUNCTIONAL DESCRIPTION

PRIMARY OXYGEN SUPPLY

This subsystem supplies oxygen for breathing and pressurizes the space suit and helmet. The oxygen is automatically fed into the suit to maintain a pressure of 3.9 psi. Approximately 1.5 pounds of gaseous oxygen is stored at between 1380 and 1440 psi in a tank nearly 6 inches in diameter and slightly more than 17 inches long. The tank is replenished from the LM oxygen supply.

OXYGEN-VENTILATING CIRCUIT

This subsystem circulates oxygen through the space-suit pressure garment and purifies recirculating oxygen. It also helps cool the astronaut by evaporating moisture that accumulates on his skin.

Oxygen entering the backpack from the suit passes through a lithium hydroxide cartridge, where chemicals trap carbon dioxide exhaled by the astronaut. It then goes through an activated-charcoal bed that removes trace contaminants, including body odors. The oxygen flow is cooled by a porous-plate sublimator, a self-regulating heat-rejection device developed by Hamilton Standard. Water in the sublimator absorbs the heat and seeps through the pores of the sublimator's sintered-nickel plates which are exposed to vacuum. The water freezes, forms an ice layer across the plates, then turns from ice to vapor. The rate of this sublimating process is governed by the amount of heat being rejected.

Excess water entering the oxygen flow, mainly from astronaut respiration and perspiration, is removed by a water separator and stored outside the bladder section of the water reservoir. A fan recirculates oxygen to the space suit at a rate of 5.5 cubic feet per minute.

Six extra lithium hydroxide cartridges are carried in the LM to replace used cartridges.

WATER TRANSPORT LOOP

This loop cools the astronaut by removing his metabolic heat and any heat that leaks into the suit from the hot lunar surface. A battery-operated pump continuously circulates 1.35 pounds of chilled water at a rate of 4 pounds per minute through a network of plastic tubing integrated in the liquid cooling garment worn under the space suit. The pack dissipates metabolic heat at an average of 1,600 Btu per hour and can handle peak rates up to 2,000 Btu.

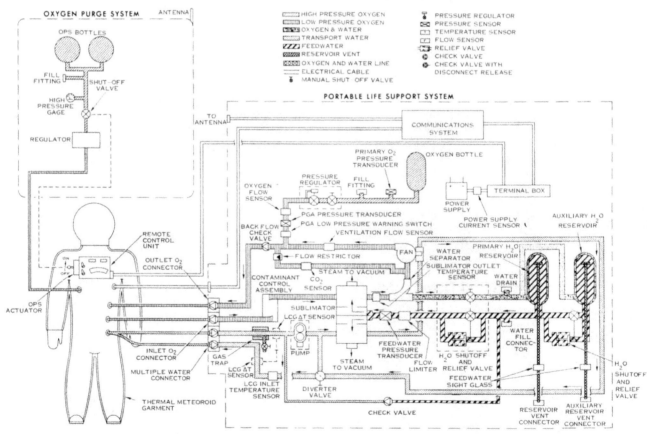

R-114A

Apollo Extravehicular Mobility Unit Schematic – Lunar Configuration

Hamilton Standard

R-115A

Front View of PLSS

R-116A

Rear View of PLSS

The sublimator that cools the oxygen flow extracts heat from the circulating water, which normally leaves the pack at 45° F. To control cooling, the astronaut uses a valve on the pack to select any one of three water temperature ranges (45° to 50°, 60° to 65°, or 75° to 80°). This valve diverts water past the sublimator.

FEEDWATER LOOP

This subsystem supplies 11.8 pounds of expendable water, stored in a rubber bladder reservoir, to the heat-rejecting porous-plate sublimator. Of this expendable feedwater, 8.5 pounds is stored in the main reservoir; an auxiliary tank holds the remaining 3.3 pounds. Suit pressure against the bladder forces water into passages between the sublimator's heat transport fluid passages and its metal plates, which are exposed to space vacuum. The ice layer formed on the porous plates during sublimation prevents the slightly pressurized water from flowing through the metal pores.

Condensed water from the oxygen-ventilating circuit is collected outside the reservoir bladder. Feedwater is replenished from the LM supply.

Refilling the bladder forces water condensed from the oxygen flow into the LM waste management system.

SPACE SUIT COMMUNICATION SYSTEM

This system, manufactured by Radio Corporation of America, provides primary and backup dual voice transmission and reception, telemetry transmission of physiological and backpack performance data, and an audible warning signal. It also regulates the voltage and electrical current of the oxygen quantity gauge and various sensors.

Operation of the communication system in the dual mode, provides crew members with uninterrupted duplex voice communications with one another, with the LM and, via the LM, with Mission Control. A dual volume control permits adjustment of receiver sound level. The transceiver control station on the LM is used as a relay station between crewmen on the lunar surface or in space. It also relays radio-telemetry data to earth monitors and to the Command Module when it is in line-of-sight of the LM.

Telemetry information is transmitted without interrupting or interfering with voice communication. Nine telemetry channels transmitted to the LM carry suit operational and environmental data - oxygen supply pressure, suit water inlet temperature, sublimator oxygen outlet temperature, suit pressure, feedwater pressure, suit water temperature rise, CO_2 partial pressure, and backpack battery current and voltage. A tenth channel transmits an electrocardiogram signal.

Indicators mounted on the remote control unit provide the astronaut with a visual warning of high oxygen usage rate, low suit pressure, low ventilation flow and low feedwater pressure. An audible tone sounds to alert the astronaut that an abnormal condition exists. Flags trip into view in the indicator windows, identifying the problem so that the astronaut can take corrective action.

OXYGEN PURGE SYSTEM

The oxygen purge system OPS, connected to the suit by a separate umbilical, is designed for backup use in the event of emergencies such as loss of suit pressure or depleted oxygen supply. However, an astronaut can use it independently as a life support chest pack during extravehicular transfer between the LM and CM spacecraft.

The system supplies either an open-loop purge flow or makeup flow directly to the suit. In both cases, it maintains suit pressure at 3.7 psi. In the full purge mode, it provides a 30-minute flow at a rate of 8.3 pounds of oxygen an hour, fulfilling breathing and cooling requirements, flushing out carbon dioxide, and defogging the helmet visor.

When used in conjunction with the Buddy Secondary Life Support System (BSLSS), the OPS flow is reduced to 4.2 pounds per hour, which permits emergency operation for up to 75 minutes.

The OPS, mounted separately on top of the backpack, is operated by a lever, attached to the pack's remote control unit. Its umbilical is attached to the suit connector that connects the suit to the LM Environmental Control Subsystem when the astronaut is inside the LM.

The purge unit weighs 35.1 pounds; is 18.4 inches long, 10 inches high, and 8 inches deep. Two spherical containers hold a total of 5.7 pounds of oxygen stored at 6,950 psi. A battery-powered, temperature-controlled heater warms the rapidly expanding oxygen to prevent subzero oxygen temperatures at the space-suit flow inlet.

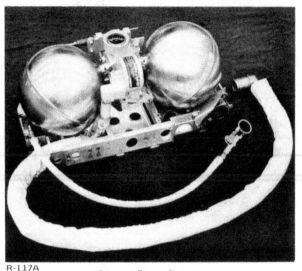

R-117A

Oxygen Purge System

BUDDY SECONDARY LIFE SUPPORT SYSTEM

The buddy system consists of two flexible hoses which feed cooling water from one astronaut's life support backpack to the other space suit if its cooling equipment fails.

Addition of the buddy system doubles the time the emergency oxygen available in the OPS can last. The oxygen purge system support pack previously supplied oxygen not only for breathing purposes and pressurizing the suit but to cool the astronaut by its high flow rate.

With the buddy system taking over the cooling function, emergency oxygen flow can be slowed down to extend its supply from 30 minutes to 60 to 90 minutes. The length of time is governed by the level of the astronaut's physical activity.

The emergency water-cooling hoses are eight and a half feet long. One hose carries water into the suit, the other out of it. A six-foot tether on the hoses snaps on the space suits and prevents the hoses from reaching their full length when they are connected, protecting the hoses and space suits against possible damage during the walk back to the LM. Water and tether lines are stowed in a pouch which is carried on the PLSS or Lunar Rover during each EVA.

In event of an emergency, the astronauts will remove the buddy system from the stowage pouch and hook the tether to the waist-restraint rings on their space suits. The astronaut whose life support pack cooling system has failed will disconnect the pack water hose. Assisting each other, the astronauts then will plug the buddy system water-cooling hoses into multiple connectors which join the life support backpack's water lines to the suits.

Information in this section relative to the Portable Life Support System was provided by Hamilton Standard, Division of United Aircraft. Complete details on the Portable Life Support System can be obtained from Hamilton Standard.

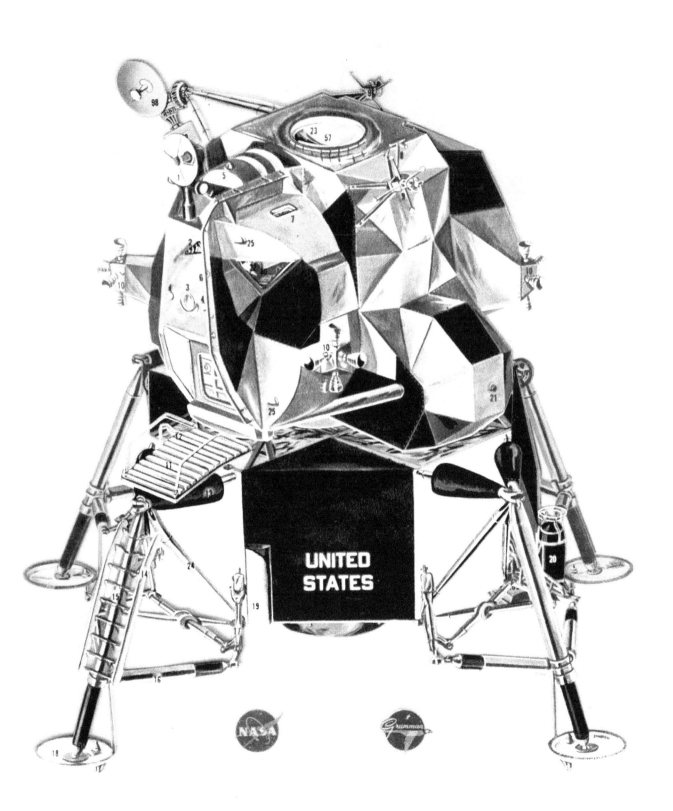

UNITED
STATES

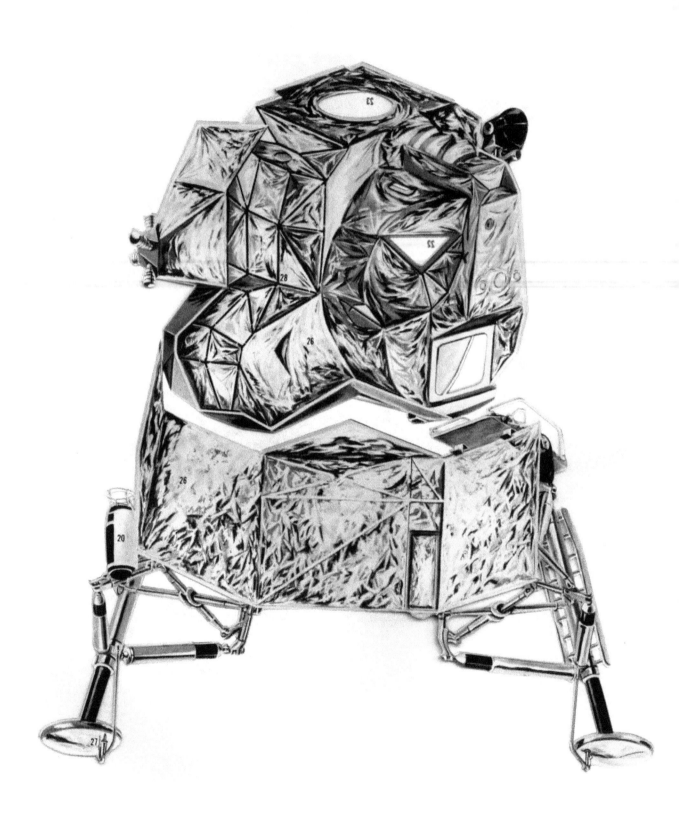

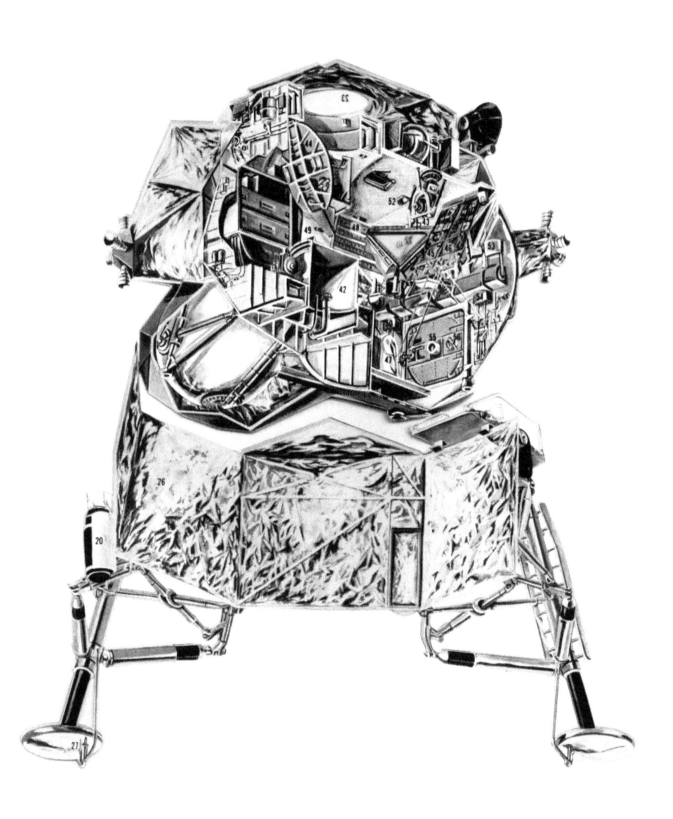

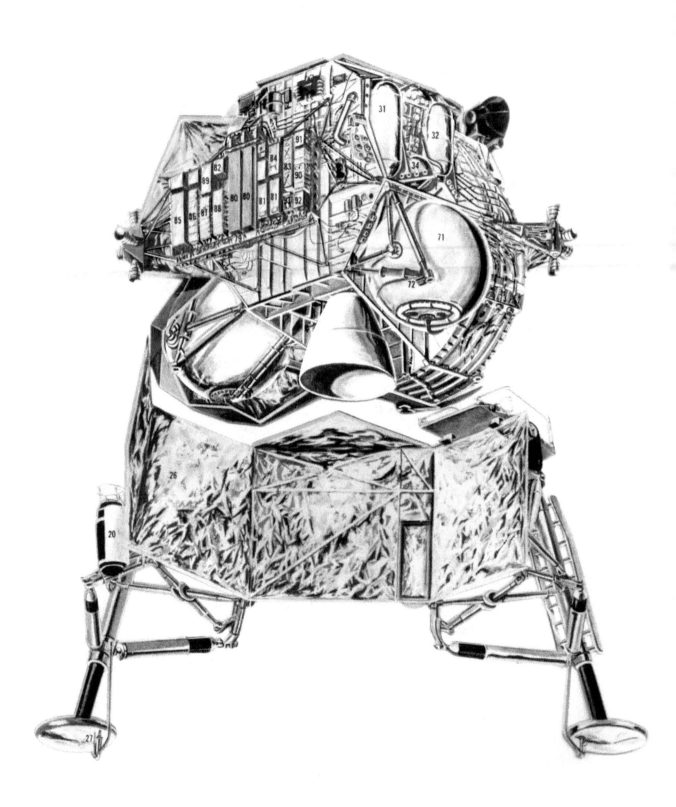

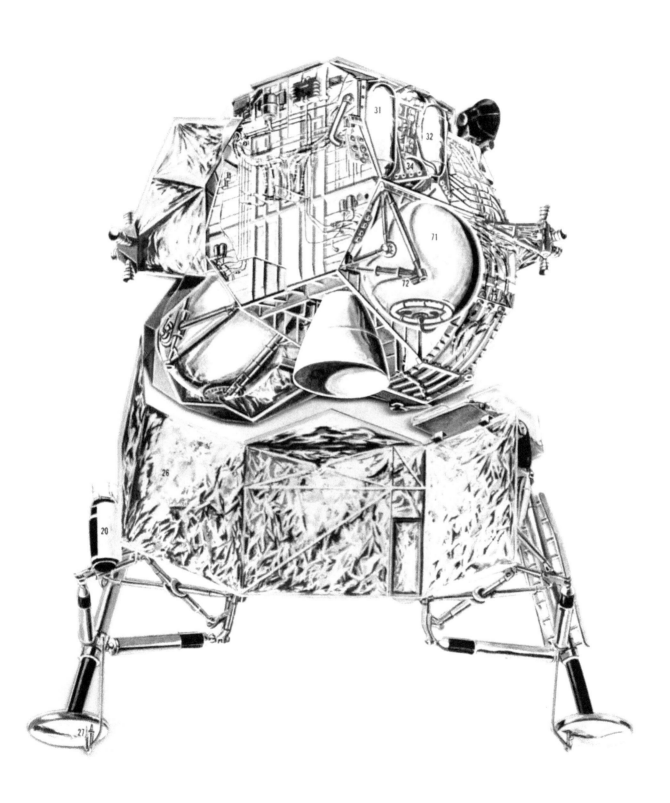

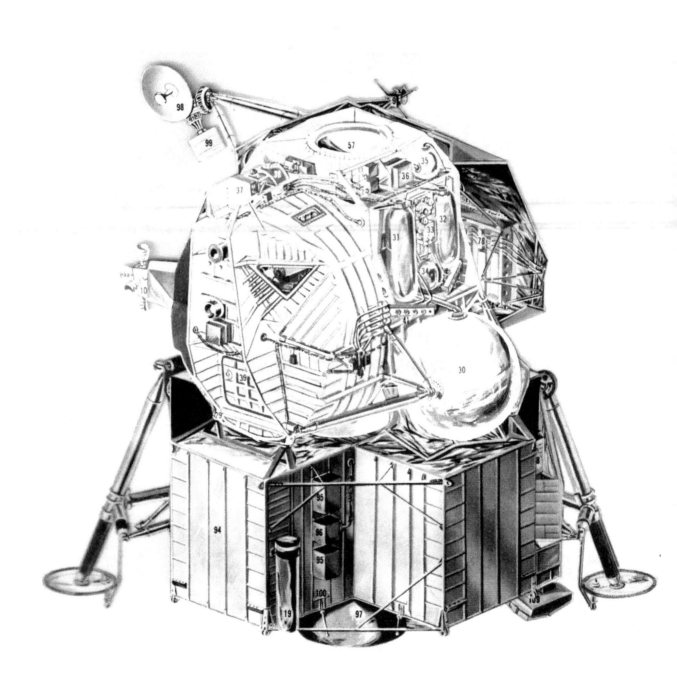

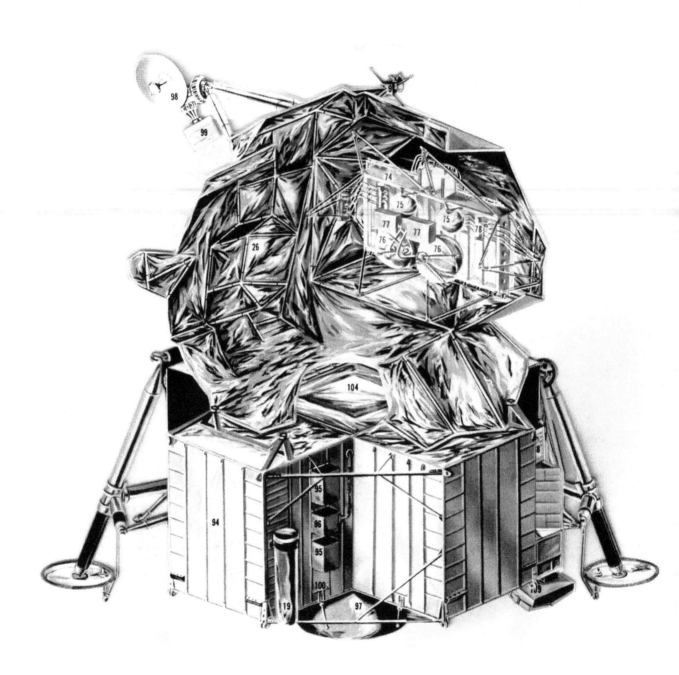

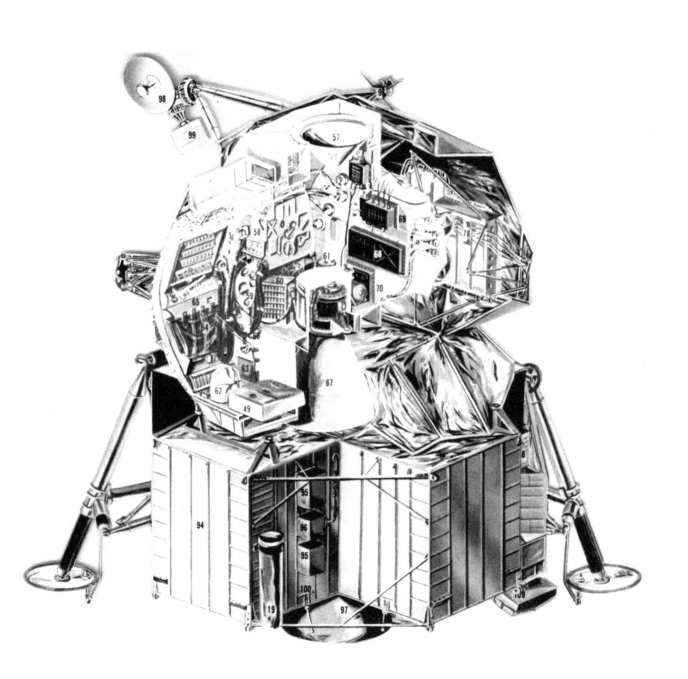

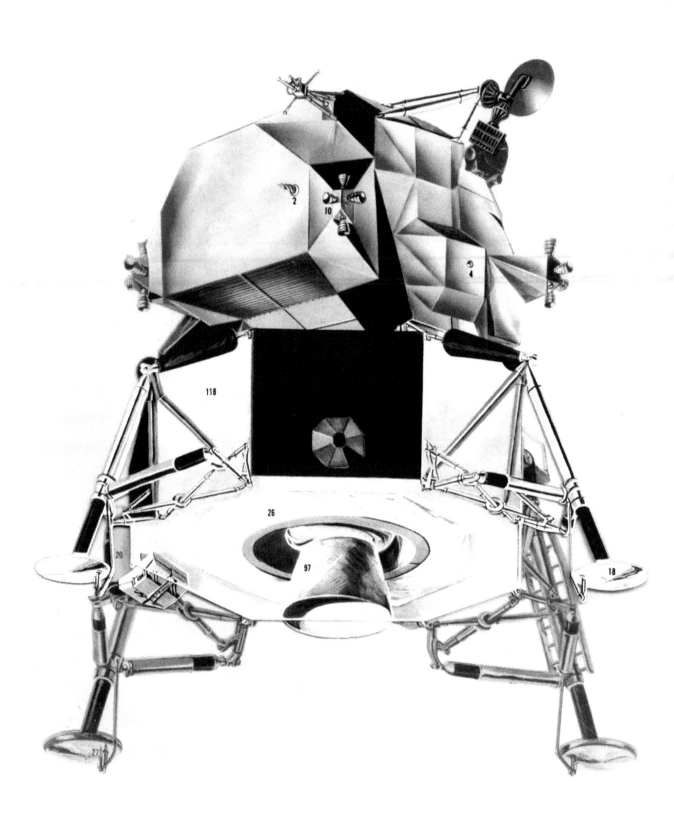

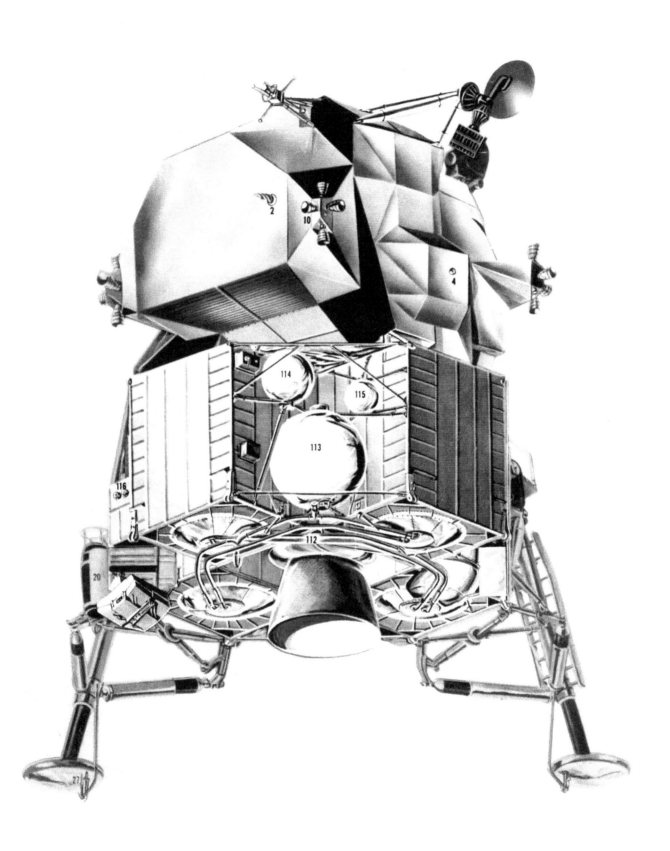

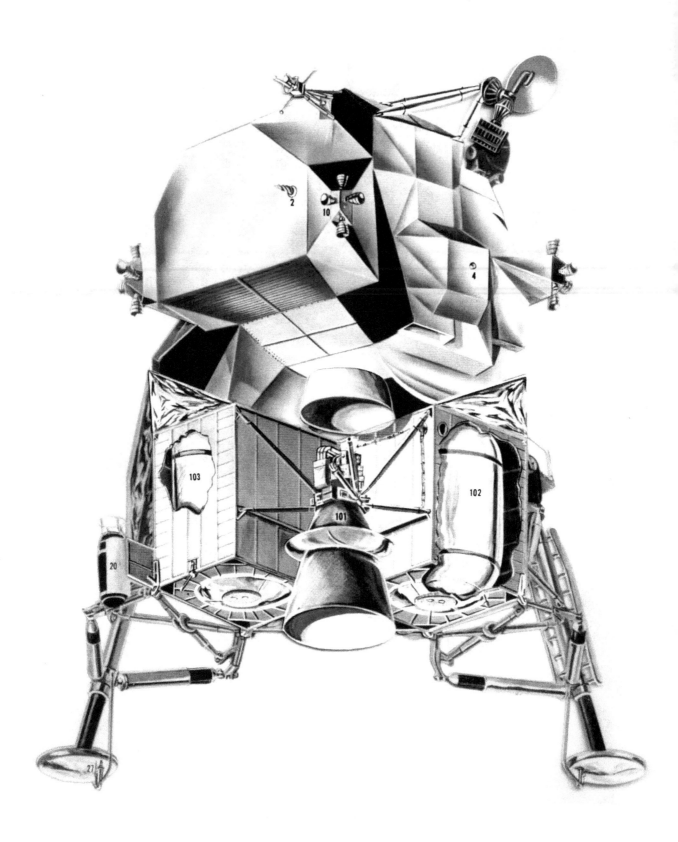

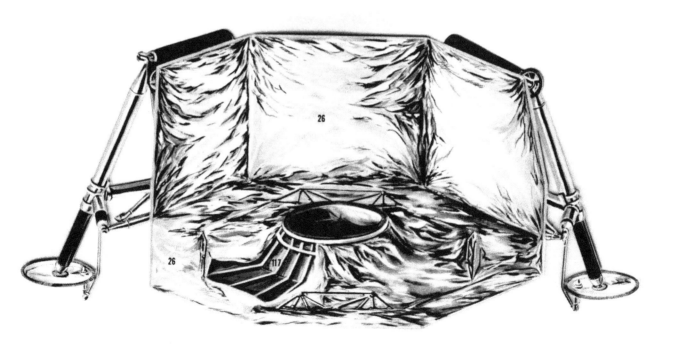

LEGEND

1. Rendezvous Radar
2. S-Band In-Flight Antenna
3. Tracking Light
4. Docking Light
5. Alignment Telescope
6. EVA Rail
7. Docking Window
8. Docking Target
9. VHF In-Flight Antenna
10. RCS Thrusters
11. Ingress/Egress Platform & Rails
12. MESA "O" Ring Release
13. Upper Outrigger Venting Shield
14. Ingress/Egress Ladder
15. Primary Shock-Absorber Strut
16. Secondary Shock-Absorber Strut
17. Deployment Truss & Down-Lock Mechanism
18. Landing Pad
19. S-Band Erectable Antenna (Lunar Surface)
20. Radioisotope Thermal Generator
21. Docking Light (Port Side)
22. Forward-Vision Window
23. LM/CM Docking Hatch
24. Outrigger Strut
25. Insulation Vent
26. Thermal Insulation Blankets
27. Lunar Surface Sensing Probe
28. Insulation Support Frame
29. Interstage Connection Points (4)
30. Ascent Fuel Tank
31. Reaction-Control Oxidizer
32. Reaction-Control Fuel
33. Helium Pressurization Unit
34. Reaction-Control Helium
35. Water Tank
36. Relay Box
37. Abort Sensor
38. Inertial Measurement Unit (IMU)
39. Ingress/Egress Hatch
40. Landing Point Designator
41. Oxidizer Service Panel
42. Ascent Engine Cover
43. Alignment Optical Telescope
44. Upper Hatch
45. Commander's Main Flight Panel
46. LM Pilot's Main Flight Panel
47. Commander's EV Visors (Stowe ')
48. Commander's Circuit Breaker Panel & Side Console
49. PLSS (Stowed)
50. Commander's Support & Restra Reel
51. Commander's Armrest & Thrust Control
52. Main Panel/Cabin Floodlights
53. LM Pilot's Armrest (Stowed)
54. LM Pilot's Support & Restraint Reel
55. Anti-Bacterial Filter Stowage
56. Cabin Relief & Dump Valve
57. Docking Drogue (Removable for Access)
58. Suit Circuit Assembly
59. Water Control Module
60. Cabin Air Recirculation Fan

60. Cabin Air Recirculation Fan
61. LiOH Canister
62. LM Pilot's EV Visor (Stowed)
63. LM Pilot's Restraint Reel
64. Crew Equipment Storage
65. LM Pilot's Console & Circuit Breaker Panel
66. Oxygen Umbilical Hoses
67. Ascent Engine (3,500 lb Thrust in Vacuum)
68. Coupling Data Unit
69. Guidance Computer & Cold Plate
70. Power Servo Assembly
71. Ascent Oxidizer Tank
72. Descent/Ascent Section Explosive Attachment
73. Interrupt Connector Assembly & Wiring
74. Aft Equipment Bay
75. Gaseous Oxygen Tank
76. Helium Tank
77. Helium Pressurization Control Modules
78. Thrust Chamber Isolation Valves
79. Electronic Replaceable Assembly Rack
80. Batteries
81. Inverter
82. Electrical Control Assembly
83. Abort Electronics Assembly
84. Attitude & Translation Control Assy
85. Rendezvous Radar Electronic Assembly
86. Signal Conditioning Electronics Replaceable Assy No. 1
87. Pulse Code Modulation & Timing Equip. Assy
88. Signal Conditioning Electronics Replaceable Assy No. 2
89. Caution & Warning Electronics Assembly
90. S-Band Transceivers
91. S-Band Power Amplifier & Diplexer
92. Signal Processor
93. VHF Transceivers & Diplexer
94. Descent Structure
95. Batteries
96. Electrical Control Assembly
97. Descent Engine Skirt
98. S-Band Steerable Antenna
99. Electronics Package
100. Landing Gear Chock Mount
101. Descent Engine Throttleable (10,000 lb Approx Thrust)
102. Descent Oxidizer Tank (Fore & Aft)
103. Descent Fuel Tank (Port & Starboard)
104. Ascent Engine Blast Deflector
105. Water Tank
106. Scientific Equipment Boxes (2)
107. Specimen Return Container Assembly (MESA)
108. Landing Radar Electronics
109. Landing Radar
110. Fuel & Electrical Line Runs

111. Fuel Lines To Descent Engine
112. Fuel Lines (Descent Engine)
113. Supercritical Helium Tank
114. Ambient Helium Tank
115. Oxygen Tank
116. Scientific Equipment Power Outlets
117. Descent Stage Skirt Structure
118. Thermal & Micrometeoroid Shield

ABBREVIATIONS USED IN LEGEND

EVA: Extravehicular Activity
VHF: Very High Frequency
MESA: Modularized Equipment Stowage Assembly
RCS: Reaction Control Subsystem
LM/CM: Lunar Module/Command Module
PLSS: Portable Life Support System

GRUMMAN

BIOGRAPHIES

NAME: Llewellyn J. Evans, Chairman of the Board and Chief Executive Officer
Grumman Aerospace Corporation, Bethpage, New York

BIRTHPLACE AND DATE: Born August 2, 1920, Unsankinko, Korea

EDUCATION: Attended Long Beach Junior College and graduated from the University of California (1942). Received a degree in Law from Harvard Law School in 1947.

MARITAL STATUS: Married to the former Georgene Hubbard of Seattle, Washington. They have a son, Llewellyn, Jr.

ORGANIZATIONS: Member of the American Bar and Federal Bar Associations, and the Harvard Law School Association, the Air Force Association and the Navy League. He is also a board member of Travelers Corporation.

MILITARY SERVICE: United States Army Air Corps from 1943 to 1945. Was Group and Squadron staff flight engineer. Was among the first crew members assigned to the B-29 bomber. Earned the Distinguished Flying Cross and Air Medal with five Oak Leaf Clusters.

BACKGROUND: Mr. Evans is a member of Sigma Alpha Epsilon. After receiving his Harvard law degree in 1947, Mr. Evans served in the Office of General Counsel with the Department of the Navy, Washington, D.C. During this period he was admitted to the Bars of the United States District Court for the District of Columbia, the United States Court of Appeals, the United States Court of Claims, and the New York State.

Mr. Evans joined Grumman in 1951 and served as Associate General Counsel until 1958 when he was appointed General Counsel. In 1960 he was elected Vice President of Grumman, and in 1963 was elected Senior Vice President. He became President and a member of the Board of Directors of Grumman Aircraft Engineering Corporation in May 1966. Mr. Evans served in that position until May 1969 when the company was restructured and renamed Grumman Corporation. At that time he became President and a Director of both Grumman Corporation and its major subsidiary, Grumman Aerospace Corporation, as well as a Director of all other Grumman subsidiaries. Early in 1971, he was elected Chairman of Grumman Aerospace Corporation.

In March 1969 Mr. Evans received the NASA Public Service Award "for his outstanding contribution as a key leader of the Government/Industry Team which made possible the exceptional success of Apollo 9, the first manned flight of the Lunar Module."

He and his family reside in Brookville, New York.

NAME: William M. Zarkowsky, President
Grumman Aerospace Corporation, Bethpage, New York

BIRTHPLACE AND DATE: Born April 22, 1919, New York, New York

EDUCATION: Graduate of New York University 1939; Graduate work at New York University, Columbia, and Adelphi University; M.S. in industrial management, Massachusetts Institute of Technology, 1958.

MARITAL STATUS: Married to the former Eleanor Frances Currie of Washington, D.C. They have four sons, Michael, graduate of Southampton College; John, a student at University of Massachusetts; Frank, a student at Southampton College, Theodore, a student at Huntington High School, and a daughter Katherine, a student at Simpson Junior High School.

ORGANIZATIONS: Mr. Zarkowsky has served on numerous engineering committees of the ASTME, AIAA, NACA, NASA and SAE. He is a member of the Board of Governors, Vice President, and National Chairman of the Society of Sloan Fellows of M.I.T., he is a member of the Executive Board of the Boy Scouts of America in Suffolk County, and also serves on the Training Committee of the Boy Scouts of America in Suffolk County.

MILITARY SERVICE: ROTC

BACKGROUND: Mr. Zarkowsky was a member of Iota Alpha & Kappa Sigma Gamma Zeta (scholastic - fraternal) at N.Y.U. After receiving his B.A.E. in 1939 he joined Grumman where he held a varied number of positions. He was named a Vice President in 1964, and a Senior Vice President in 1970. He was named Chairman of the Executive Operations Board in 1970 and has been a Director of Grumman Aerospace Corporation since May of 1969. Early in 1971 he became President of Grumman Ēcosystems Corporation and also President of Grumman Aerospace Corporation.

Mr. Zarkowsky and his family reside in Huntington, New York.

NAME: Joseph G. Gavin, Jr., Sr. Vice President
Grumman Aerospace Corporation, Bethpage, New York

BIRTHPLACE AND DATE: Born September 18, 1920, Somerville, Massachusetts

EDUCATION: Graduate of Massachusetts Institute of Technology (1942).
Received a S.B., S.M. in Aeronautical Engineering

MARITAL STATUS: Married to the former Dorothy Dunklee of Brattleboro,
Vermont. They have two sons, Joseph G. Gavin III and Donald Lewis
Gavin, a student at Wesleyan University, Connecticut. A daughter, Tay
Gavin Erickson is married.

ORGANIZATIONS: Member of Educational Council, Massachusetts Institute of Technology, American
Institute of Aeronautics and Astronautics and American Astronautical Society.

MILITARY SERVICE: United States Navy from 1942 to 1946.

BACKGROUND: Mr. Gavin was in the Honors Group of his graduating class at MIT. After his discharge from
the Navy, he began his career at Grumman as a design engineer. From the Preliminary Design group
Mr. Gavin was assigned as Project Engineer on the F9F-6 in 1950 and in 1952 as Project Engineer on
the F11F Program. In 1956 he was assigned as Chief Experimental Project Engineer until 1957 when
he was made Chief Missile and Space Engineer. In 1962 Mr. Gavin was promoted to Vice President,
Director of LM Program and in 1968 assumed the responsibility for all Grumman's space programs.
Mr. Gavin has served as President of the Harborfields Central School District #6 Board of Education.
In 1971 he received the NASA Distinguished Public Service Medal. Mr. Gavin also serves on the
Board of Directors of Grumman Aerospace Corporation.

He and his family reside in Huntington, New York.

NAME: Dr. Ralph H. Tripp, Vice President - LM Program Director
Grumman Aerospace Corporation, Bethpage, New York

BIRTHPLACE AND DATE: Born Denton, Montana, March 11, 1915

EDUCATION: Received his BA from Drake University in 1937, and his MS (1939) and PhD (1942) in Applied Mathematics and Theoretical and Applied Mechanics from Iowa State College.

MARITAL STATUS: Married the former LaVone Semingson. They have three daughters, Virginia, Haillie, and Roberta.

ORGANIZATIONS: Fellow of Instrument Society of America (President, 1961), Associate Fellow of AIAA, Member of AAS, National Space Club, American Management Association.

BACKGROUND: Dr. Tripp was a teacher at Iowa State College until he received his PhD in 1942. He then joined Grumman working in the Structures group in charge of vibration and flutter. In 1948 he became the head of the Structural Research Group. Later that year he formed the Research Department and became the Department Head. In 1949 he became the head of the Instrumentation Department. In 1958 he became the Assistant Director of Flight Test. He was appointed Director of the Orbiting Astronomical Observatory (OAO) in 1962. He directed the OAO Program until February 1968 at which time he became LM Program Director.

He and his family reside in Cold Spring Harbor, New York.

GRUMMAN

A BRIEF HISTORY OF
GRUMMAN AIRCRAFT ENGINEERING CORPORATION

Grumman has come a long way since it opened shop in a rented garage in 1930. Its six founders and fifteen employees, within a year, fulfilled their first government contract: delivery of two amphibious aircraft pontoons. Today with more than 25,000 employees in 35 Long Island plants and 25 field locations, the Corporation is involved in research, development and production programs that encompass aircraft, spacecraft, support equipment, land vehicles, surface vessels, and submersibles.

The story of Grumman military aircraft begins in 1933 with the development of the FF-1 (a Navy biplane fighter) and proceeds to the Navy *Intruders,* the Army *Mohawks* and the current Navy F-14 *Tomcat.*

The Corporation produced the *Denison* hydrofoil boat for the U. S. Maritime Administration, the *Dolphin* hydrofoil for commercial service and the *PG(H) Flagstaff,* a military hydrofoil. For undersea research, a Grumman research submersible, the *Ben Franklin,* was designed and built for the historic Gulf Stream Drift Mission.

In commercial aviation, Grumman is producing the *Ag-Cat* for crop dusting and spraying, as well as the fan-jet *Gulfstream II* corporate transport, the follow-on to the *Gulfstream I,* of which 200 were produced.

With the award of a contract from the Department of Transportation, Grumman moved ahead in a new field of passenger conveyance, the *Tracked Air Cushion Vehicle* (TACRV). Wind tunnel testing has been conducted and testing is being scheduled at the Federal High Speed Ground Transportation Test Center in Pueblo, Colorado for the Grumman TACRV.

But this takes us ahead of the Grumman story.

Early in its existence, the Corporation developed a reputation for excellence in design and manufacture of aircraft; qualities that came to the forefront during the Second World War. Not only did Grumman build and deliver more than 17,000 combat planes during that period, but it won five Navy "E" production awards, received a Presidential Medal of Merit, and established an unequalled military production record (more than 600 *Hellcats* in just one month from a single plant). Grumman *Hellcats, Wildcats* and *Avengers* accounted for about two-thirds of the enemy aircraft destroyed in the Pacific Theater.

Grumman also has an admirable record in the commercial field. Beginning in 1936 with the appearance of the amphibious *Goose* and through the present-day success of the *Gulfstream II* - a twin-jet corporate transport - Grumman's commercial craft have established worldwide reputation for service and durability.

The coming of the space age produced the greatest period of expansion at Grumman since World War II. It also established Grumman and its geographic location on Long Island, as one of the prime technological sources for support of the United States in space.

In 1960, the National Aeronautics and Space Administration (NASA) awarded Grumman its first major aerospace contract, the development of the Orbiting Astronomical Observatory (OAO) series of spacecraft.

The most far reaching aerospace contract, however, was awarded in 1962, when NASA selected Grumman to develop the *Apollo Lunar Module* (LM). Grumman is presently under contract to NASA to study the *High Energy Astronomy Observatory* (HEAO) and the *Space Shuttle*.

In 1969, a structural reorganization of the Corporation was accomplished, insuring Grumman continued growth within the aerospace industry as well as into many other diverse industries. For forty years the company had been known as Grumman Aircraft Engineering Corporation. The words "Aircraft Engineering were dropped from the title. A parent organization was formed - Grumman Corporation - which now directs the operations of the many subsidiaries: Grumman Aerospace Corporation, Grumman Allied Industries, Inc., Grumman Data Systems Corporation, Grumman Ecosystems Corporation, Grumman International, Inc., and Montauk Aero Corporation.

Each of these subsidiaries with its individual marketing targets and leadership teams are now managerially and operationally developing their own business objectives.

A BRIEF HISTORY OF THE
APOLLO LUNAR MODULE

It has been nearly ten years since Grumman undertook the responsibility of designing, developing, and manufacturing the Apollo Lunar Module for the National Aeronautics and Space Administration.

In 1958, after many years as a producer of aircraft for the United States Navy, Grumman began studies on manned space flight programs. The Corporation submitted a spacecraft proposal for Project Mercury, but, primarily because of prior Navy aircraft production commitments, the bid was unsuccessful.

In early 1960, Grumman submitted a preliminary study on Project Apollo. At that time, the Earth Orbit Rendezvous (EOR) theory for a lunar landing was most prevalent. When NASA held a competition, Grumman and International Telephone and Telegraph presented results of their combined studies. They were unsuccessful in this bid for the project.

By 1961, Grumman, convinced they had the ability to build dependable spacecraft, submitted the results of in-house studies to NASA. In October of that year, NASA held a competition for Apollo hardware. Again, Grumman, this time in conjunction with General Electric, was unsuccessful.

In mid-1962, thoroughly convinced that Lunar Orbital Rendezvous (LOR) was the best method to effect a lunar landing, Grumman launched a feasibility study on LOR. NASA then asked for proposals involving use of the LOR concept and the Lunar Excursion Module. Grumman submitted its proposal in September 1962, with RCA as principal subcontractor. NASA Administrator James E. Webb, now retired, emphasized at the time that only since July had NASA committed itself to "lunar orbit rendezvous" using the advanced Saturn booster. More than a million man-hours had gone into studies of how to get men to the moon and back.

On November 7, 1962, NASA issued the following news release:
    "Grumman Aircraft Engineering Corporation, New York, today was selected to build Project Apollo Lunar Excursion Module — a spacecraft in which Americans will land on the moon and return to a moon orbiting mother craft for the journey back to earth".

By November 18, 1962, a team of Grumman engineers was in Houston working with NASA, even though the contract still was being negotiated.

Items listed in the news release included:
- A three-stage Saturn with first-stage thrust of 7.5 million pounds
- A 5-ton Command Module (now 6½ tons)
- A 25-ton Service Module (now 27½ tons)
- A 12-ton Lunar Excursion Module (now 18 tons)

The NASA release went on to say, "LEM will look something like the cab of a two-man helicopter, measuring 10 feet in diameter and standing about 15 feet tall on its skid-type legs". (It now measures 31 feet, legs extended, and stands 23 feet tall.)

Many facets of the module have changed from the initial drawings and proposals made in 1960, even its name. NASA dropped the "E" (for "Excursion") in LEM in 1967.

To date, Grumman has received approximately $1.8 billion on its cost incentive contract with the National Aeronautics and Space Administration for design, development, and manufacture of the Apollo Lunar Module. Although the original contract called for 15 LM flight vehicles, 10 Lunar Module Test Articles (LTA's) and two simulators, reductions in the Apollo Program will leave some completed and partially completed LM's unused. At its peak, Grumman employed more than 8,000 people on the Apollo Program at Bethpage, New York; White Sands, New Mexico; Cape Kennedy, Florida; and Houston, Texas. There are now approximately 3,000 people assigned to the LM Program.

R151   *1962*

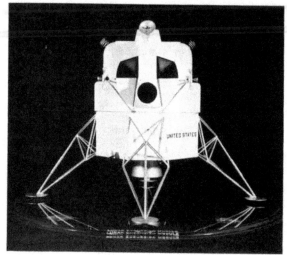

R152   *1963*

R153   *1965*

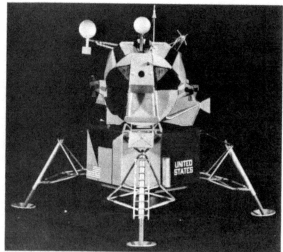

R154   *1971*

GRUMMAN

APOLLO LAUNCH HISTORY

APOLLO	SATURN	COMMAND MODULE	LUNAR MODULE	MISSION	PRIME CREW	LAUNCH DATE	FLIGHT TIME (Hrs: Min: Sec:)	REVOLUTIONS	SPLASH-DOWN DATE
4	501	017	CTA-10R	Earth Orbit	Unmanned	11-9-67			Reentry 11-9-67
5	204		1	Earth Orbit	Unmanned	1-22-68			Reentry A/S 1-24-68 D/S 2-12-68
6	502	020	LTA-2R	Earth Orbit	Unmanned	4-4-68			4-4-68
7	205	101		Earth Orbit	W.H. Schirra D. Eisele W. Cunningham	10-11-68	260:8:45	163	10-22-68
8	503	103	LTA-B	Lunar Orbit	F. Borman J.L. Lovell, Jr. W. Anders	12-21-68	147:00:41	10	12-27-68
9	504	104 Spider	3 Gum Drop	Earth Orbit	J.A. McDivitt D.R. Scott R.L. Scweickart	3-3-69	241:00:53	151	3-13-69
10	505	106 Charlie Brown	4 Snoopy	Lunar Orbit	T.P. Stafford J.W. Young E.A. Cernan	5-18-70	192:03:23	31	5-26-69
11	506	107 Columbia	5 Eagle	Lunar Landing (7-20-69)	N.A. Armstrong M. Collins E.E. Aldrin, Jr.	7-16-69	195:18:35	30	7-24-69
12	507	108 Yankee Clipper	6 Intrepid	Lunar Landing (1-19-69)	C. Conrad, Jr. R.F. Gordon, Jr. A.L. Bean	11-14-69	244:36:25	45	11-24-69
13	508	109 Odyssey	7 Aquarius	Lunar Landing (Aborted)	J.A. Lovell, Jr. J.L. Swigert, Jr. F.W. Haise, Jr.	4-11-70	142:54:41		4-17-70
14	509	110 Kitty Hawk	8 Antares	Lunar Landing	A.B. Shepard, Jr. S.A. Rossa E.D. Mitchell	1-30-71	216:01:59	34	2-9-71
15	510	112 Endeavour	10 Falcon	Lunar Landing	D.R. Scott A.M. Worden J.B. Irwin				
16	511	113	11	Lunar Landing	J.W. Young T.K. Mattingly C.M. Duke, Jr.				
17	512	114	12	Lunar Landing					

LM MANUFACTURING

The ascent stage of the Apollo Lunar Module (LM) is the control center and manned portion of the space vehicle. Its three main sections are the crew compartment, midsection, and aft equipment bay and tank section. The crew compartment and midsection make up the cabin. The ascent stage structure consists of the following subassemblies: front face, cabin skin, midsection, and aft equipment bay. The cabin skin subassembly is fabricated from formed chem-milled skin panels that are welded and mechanically fastened.

R-122

The front face of the ascent stage is fabricated from chem-milled skin panels that are welded and mechanically fastened. Sealing the mechanical joints, trimming the forward face contour, and adding formed longerons and stringers complete the operations for this assembly.

R-123

The midsection consists of two machined bulkheads, an upper deck tunnel weldment, a lower engine deck weldment, and chem-milled skins.

GRUMMAN

T STAGE
R-124

The front face assembly and cabin skin subassembly are mechanically joined with the midsection and are sealed to form the cabin pressure shell of the ascent stage.

R-125

Cold rails, chem-milled beams, struts, and machined fittings comprise the major structural components in the aft equipment bay.

The descent stage is the unmanned portion of the LM. It consists primarily of machined parts and chem-milled panel/stiffener assemblies that are mechanically fastened. Compartments formed by the structural arrangement house the descent engine, and propellant, helium, oxygen, and water tanks.

R-126

Fabrication of the descent stage begins with the joining of the machined "picture frames" and the chem-milled panel/stiffener assemblies to form the engine compartment.

R-127

After the outrigger bulkhead assemblies are attached to the engine compartment with machined cap strips, the eight remaining panel/stiffener assemblies are added.

The cantilever-type landing gear is attached externally to the descent stage and folds inward to fit within the shroud of the Saturn V aerodynamic shell. The landing gear consists of four sets of legs connected to outriggers that extend from the ends of the descent stage structural beams.

R-128

With the addition of the upper and lower machined decks and the machined interstage fittings, the completed descent stage structure is moved to the clean room facility.

Each landing gear consists of a primary strut and foot pad, two secondary struts, an uplock assembly, two deployment and downlock mechanisms, a truss assembly, and a lunar-surface sensing probe. A ladder is affixed to the forward leg assembly. The struts are machined aluminum with machined fittings mechanically attached at the ends.

R-129

GRUMMAN

The Descent Propulsion Section consists of two fuel and two oxidizer tanks centered about a deep-throttling ablative rocket engine which has restart capabilities.

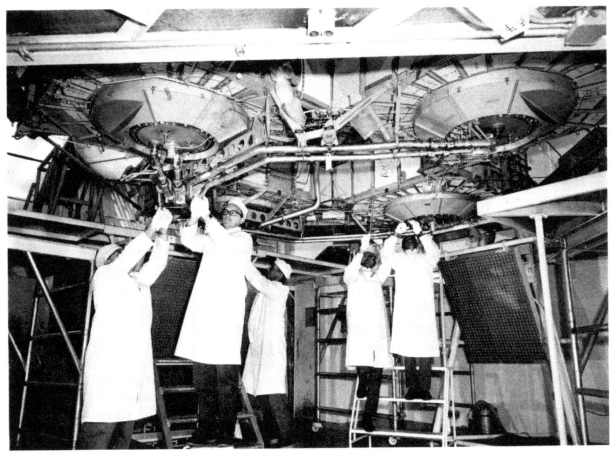

R-130

After the descent stage has been moved to the clean room facility, interconnecting gas and liquid balance lines for like tanks are installed.

The Ascent Propulsion Section uses a fixed, constant-thrust rocket engine. The section includes the associated ambient helium pressurization and propellant supply components.

R-131

With the installation of the various electrical and electronics components and associated wiring, the two stages of the LM are tested and checked out separately.

Two main propellant tanks are used; one for fuel, the other for oxidizer. The tanks are installed on either side of the ascent stage structure.

The ascent and descent stages are then mated and further checks are made on the entire spacecraft.

R-132

R-133

Although strict cleanliness procedures are followed while the LM is under construction and test, one last clean and rotate check is made. Loose material overlooked by the quality control teams will be dislodged and removed during this process.

When all components of the LM subsystems have been verified, the installation of thermal blankets and micrometeoroid shielding begins. The spacecraft is now ready for Final Engineering and Acceptance Testing.

R-134

R-135

Prior to shipment, the stages of the Lunar Module are separated and a landing gear deployment check is made. The landing gear is then removed prior to the LM being put into a protective container.

R-136

The Lunar Module ascent stage is then prepared for shipment. Technicians verify that all components are properly secured.

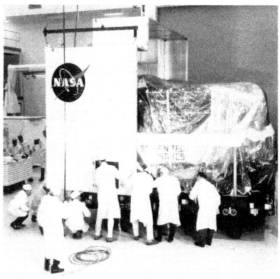

R-137

The stages are put into protective containers. When the entire stage has been encased, dry nitrogen is pumped into the container and maintained at positive pressure during the flight to NASA Kennedy Space Center.

GRUMMAN

R-138

The separately packaged Lunar Module stages are placed aboard the Super Guppy aircraft for the flight to NASA Kennedy Space Center.

GLOSSARY

Ablation — Shedding of excessive heat from a surface by vaporization or melting of specially designed coating materials (ablative material). The Apollo heat shield disperses reentry heat in this manner; the combustion chamber and nozzles of the descent and ascent engines are also ablatively cooled.

Abort — To cut short a launch or mission because of equipment failure or other problems.

Acceleration — Rate of gain in velocity.

Accelerometer — "Speedometer" in spacecraft control system; a device that measures velocity changes along all three axes and sends signals to the guidance computer, displays, etc.

Accumulator — Storage device, such as hydraulic system apparatus, which stores fluid under pressure, or a computer device, which stores a continuously higher sum as it adds incoming numbers to that sum.

Acquisition — Process of locating an orbiting spacecraft to begin tracking or gathering telemetry data.

Activated Charcoal — Substance used in atmospheric revitalization section to remove odors from recirculating cabin and suit oxygen supply.

Actuators — Devices that transform an electrical signal into a mechanical motion, using hydraulic or pneumatic power.

Aerozine — Liquid fuel used in ascent and descent engines; half unsymmetrical dimethylhydrazine and half hydrazine. It is storable, and hypergolic in the presence of nitrogen tetroxide.

Aft Equipment Bay — Unpressurized area in the ascent stage for electronic equipment, batteries, oxygen supply, and cooling equipment.

Ambient — Denotes "normal" environmental conditions such as pressure or temperature. (LM cabin ambient is 4.5 psi at 70° F.)

Analog Computer — Computer that operates on the principal of measuring (linear lengths, voltages, etc.), as distinguished from counting. An analog computer in the Lunar Module converts water-quantity measurements into a form suitable for display.

Apollo — NASA's manned lunar landing program, and the spacecraft built to achieve it. Originally, the Greek god of light.

Ascent Engine — The 3,500-pound thrust engine in the ascent stage, used for launch from the moon's surface and orbital adjustment, or prelanding abort.

Ascent Stage — Upper portion of the Lunar Module; houses crew, controls, and ascent engine. It returns the crew to the Command Module in lunar orbit.

Atmosphere Revitalization — Replenishing, cleaning, dehumidifying, deodorizing, and cooling or heating the air in the Lunar Module atmospheric system.

Attitude — Position or orientation of the spacecraft as determined by the inclination of its axes to some reference line or plane.

Attitude Control Mode — One of two major modes with which system spacecraft attitude is maintained.

Audio Center — Portion of the Lunar Module Communications Subsystem, including earphone and microphone controls, voice-operated relay controls, and the voice recorder.

Axis — Any of three straight lines about which a spacecraft rotates; one of a set of reference lines for a coordinate system.

Backpack — Self-contained portable life support system.

Backup — Item or system available as replacement for one that fails; an astronaut or astronaut crew trained to replace the prime pilot(s) in the event of illness or death.

Biosensors — Small devices attached to crewmembers to sense heartbeat and respiration rate.

Bipropellant — Using two propellants (fuel and oxidizer), which are fed separately into the combustion chamber.

Bit — Abbreviation of binary digit; smallest unit of computer-coded information, carried by a single digit of binary notation.

Blowout Disk — Thin metal diaphragm used as a safety device to relieve excessive gas pressure. (See Burst Diaphragm.)

Burn — The firing of engines or to fire them. Burn time is the length of the thrusting period.

Burst Diaphragm — Thin metal disk, which ruptures at a predetermined point to relieve excessive pressure. (Also Burst Disk.)

Caution and Warning System — System that monitors spacecraft subsystems and causes master alarm lights to go on and a warning tone to be initiated if malfunctions or critical conditions are detected.

Cavitation — Rapid formation and collapse of vapor pockets in a flowing liquid at low pressure; causes structural damage to rocket components; formation of partial vacuum in a pump, such as fuel pump.

Celestial — Pertaining to the stars. Celestial mechanics pertains to the motion of bodies in gravitational fields. Celestial navigation is onboard navigation, using stars for reference.

Center of Gravity — Central point of a body with regard to the distribution of its mass; the point at which its weight is centered.

Chamber Pressure — Pressure in rocket-engine combustion chamber.

Cold Plates — Equipment mounting surfaces made of sealed parallel flat plates with coolant passages. Water-glycol, circulating through the passages, removes heat from the mounted equipment.

Cold Rails — Same as cold plates except formed of channel- and tube-type extrusions.

Command — A pulse or signal initiating a step or sequence.

Command and Service Module — Combined Command Module and Service Module, which remains in lunar orbit after the Lunar Module descends to the moon. The Command and Service Modules are not separated from each other until shortly before reentry to earth's atmosphere.

Command Module — Apollo spacecraft's control center and living quarters for most of the lunar voyage. A cone 12 feet high by 12 feet 10 inches at the base, it is the only part of the spacecraft that will reenter the earth's atmosphere. It provides about 70 feet of living area per man (nearly double what was available in the Gemini spacecraft), weighs about 12,500 pounds at launch, and is covered by an all-over heat shield.

Command Module Pilot — Title of a member of the flight crew, occupying the center couch in the Command Module; the one crewmember who will not set foot on the lunar surface during a lunar landing mission, but will remain in the Command Module in lunar orbit. He is the expert on Command Module systems, the primary navigator during the trip, and the second man in seniority.

Commander — See Spacecraft Commander.

Comparator — Electronic circuit that compares one set of data with another.

Configuration — Shape; figure or pattern formed by relative position of various things.

Constant Wear Garment — Astronaut flight "underwear" or "shirtsleeves", worn under pressure suit; replaced by liquid-cooled garment during lunar exploration. (Pressure suits will be removed for part of the flight.)

Converter — A unit that changes the language of information from one form to another.

Coupling Data Unit — Assembly of electromagnetic transducers and gears, and displays, to present coordinated data from the guidance and navigation equipment; couples analog signals of IMU and optics, and converts to digital signals for guidance computer.

Crewman Optical Alignment Sight — Range-finder type of device used to help astronauts align the Command or Lunar Modules with each other during docking.

Cryogenic — Supercold, -195° C or less; refers to fuels or oxidizers that are liquid only at very low temperatures.

Daily Metabolic Requirement — For a man of 154 pounds, about 2 pounds of oxygen, 5 pounds of water, and 1 pound of solid food a day. He produces waste products of about 2 pounds of carbon dioxide and 6 pounds of water, urea, minerals, and solids. The intake rate is used as a rule of thumb in loading consumables, such as water and oxygen, for space flight.

Damping — Restraining; slowing down or stopping.

Deadband — In a control system, the range of values through which a measure can be varied without an effective response; the "play" in the control.

Decibel — Measure of sound. The human ear has a comfortable range of 1 to 130 decibels, 1 being the faintest sound a human can hear. Sounds over 130 decibels cause pain.

Delta P (ΔP) — Differential Pressure.

Delta V (ΔV) — Velocity change.

Descent Engine — Gimbaled engine on the descent stage; may be throttled to any thrust power between 1,050 and 10,500 pounds, operated automatically by the Guidance, Navigation and Control Subsystem or manually by the LM crew. It is used to descend from the Command Module (in lunar orbit) to the surface of the moon.

Descent Stage — Lower portion of the Lunar Module, containing descent engine and propellant tanks, landing gear, and storage sections. It serves as a launching platform for the ascent stage when the crew lifts off from the moon. It remains on the lunar surface.

Digital Computer — Computer that uses the principal of counting as opposed to measuring. (See Analog Computer.)

Destructive Readout — Readout of data stored in a computer memory that results in the data being erased.

Diplexer — Device that permits an antenna system to be used simultaneously by two transmitters.

Display — Visual presentation of data, usually from sensors or measuring devices, processed through a conditioning system.

Docking — Closing and mating together two spacecraft, following rendezvous.

Docking Drogue — Latching device, in the Lunar Module, into which the Command Module probe is pushed during docking; may be mounted or removed from the transfer tunnel by the crew.

Docking Latches — Four semiautomatic and eight manual latches to hold the Command and Lunar Modules firmly together when docked; the semiautomatic latches operate the docking probe retraction mechanism when engaged.

Docking Probe — Three-legged extendable device attached to the docking ring on the Command Module. It engages a drogue on the Lunar Module; may be mounted or removed from the transfer tunnel by the crew.

Docking Ring — Aluminum structure just forward of the top Command Module hatch; contain the Lunar/ Command Module seals and a pyro charge and serves as a mounting point for the docking probe and latches.

Docking System — Docking ring, probe, drogue, latches, crewman optical alignment sight, and tubular member device to be used in docking and crew transfer.

Docking Tunnel — Tunnel through which crew transfers between Lunar and Command Modules; located half in the nose of the Command Module and half in the top of the Lunar Module. It contains mounting points for the probe and drogue.

Doppler Effect — Apparent change in the frequency of sound waves (pitch), light, and radio and radar waves when the distance between the source and the observer or receiver is changing.

Doppler Principle — A principle of physics that states that, as the distance between a source of constant vibrations and an observer diminishes or increases, the frequencies appear to be greater or less.

Dosimeter — Device worn on right side of astronaut helmets and in pockets of the constant wear garment, for measuring and recording the amount of radiation to which the astronaut is exposed.

Downlink — Part of the communications link that receives, processes, and displays data from the spacecraft.

Egress — As a verb, to exit the spacecraft, as an adjective, describes the exit hatchway, procedures for exiting, etc.

Event Timer — Instrument that times an event and displays time taken to perform it.

Exploding Bridgewire — Metal wire that disintegrates at high temperature produced by a large electrical pulse; used for initiating stage retro-rockets, separation systems, etc.

Explosive Bolts — Bolts that attach the ascent and descent stages; surrounded by an explosive charge which is actuated by an electrical impulse when stage separation is desired.

Explosive Bridgewire — Wire which heats to a high temperature and burns, thus igniting a charge.

Extravehicular — Indicates that an element, such as an antenna, is located outside the vehicle.

Extravehicular Mobility Unit — Space suit (including water-cooled undergarment, pressure suit, integral thermal micrometeoroid garment, boots, gloves, helmet, visors, and portable life support system) used during lunar stay.

Flight Director Attitude Indicator — Device on control panel, which displays spacecraft attitude, attitude error, and rate of attitude change. Signals are supplied to the indicator by the Guidance, Navigation, and Control subsystem.

Gimbal — Frame with two or three mutually perpendicular and intersecting axes of rotation on which an engine or other device can be mounted and which allows it to move or swivel in two or three directions.

Guidance Computer — Digital computer, using erasable and fixed memory; computes deviations from required flight path and calculates attitude and thrust commands to correct them.

Guidance System — A system which measures and evaluates flight information, correlates this with target data, converts the result into the conditions necessary to achieve the desired flight path, and communicates this data in the form of commands to the flight control system.

Gyroscope — Device that uses angular momentum of a spinning rotor to sense angular motion of its base about one or two axes at right angles to the spin axis.

Heat Exchanger — Device for transferring heat from one fluid to another without mixing the fluids. In the Lunar Module, unwanted heat is absorbed by a water-glycol mixture and transported to sublimators.

Heat Sink — A contrivance for the absorption or transfer of heat away from a critical part or parts. (See Cold Plates and Cold Rails.)

Helium — Gas used to pressurize propellant tanks and force propellant into feed lines.

Hertz — One cycle per second.

Hover and Translation Maneuver — Maneuver of the Lunar Module, during lunar descent, to remain at a constant attitude above the moon's surface while moving laterally above the landing area.

Hypergolic — Self-igniting. Hypergolic fuel ignites spontaneously upon contact with its oxidizer, thereby eliminating the need for an ignition system.

Inertia — Tendency of an object at rest to remain at rest and of an object in motion to remain in motion in the same direction and at the same speed until gravity or some other force slows or stops it.

Inertial Guidance — Navigation system, using gyroscopic devices and a computer, that functions without external information. It automatically adjusts the vehicle to a predetermined flight path. Basically, it knows where its going and where it is by knowing where it came from and how it got there.

Inertial Measurement Unit — Main unit of the inertial guidance system; consists of a stable platform (inertial platform) that contains three inertial reference integrating gyros, three integrating accelerometers, and three angular differentiating accelerometers. It senses attitude changes or acceleration of the spacecraft.

Inertial Reference Integrating Gyro — Single-degree-of-freedom gyro that senses displacement of the stable platform on which it is mounted and generates signals accordingly.

Ingress — As a verb, getting into the spacecraft; as an noun, the entrance hatch, etc.

Injection — Introduction of fuel and oxidizer into the combustion chamber of an engine. The device that does this is an injector.

Integrated Thermal Micrometeoroid Garment — Bulky outer garmet covering pressure suit and backpack; worn for protection against extremes of heat and micrometeoroids by crewmen exploring the lunar surface. Made of lightweight felt and aluminized mylar, it limits the heat leak into the suit to about 250 Btu per hour.

Interface — Common boundary between one part of the Lunar Module or its subsystems and another; the place where two parts of a subsystem meet.

Integrating Accelerometer — Mechanical and electrical device that measures the force of acceleration along the longitudinal axis, records velocity, and measures the distance traveled.

Interstage — Between Lunar Module stages.

Inverter — Device for converting direct current produced by the Lunar Module batteries to alternating current.

Liquid-Cooled Garment — Undergarment worn beneath the pressure suit during exploration on the lunar surface. A small electrical pump in the backpack circulates water through tiny capillary tubes throughout the garment, and a heat exchanger in the backpack cools it each cycle.

Lithium Hydroxide — Substance used to remove exhaled carbon dioxide from the oxygen atmosphere of cabin and suits before recirculating the oxygen; carried in 4-pound canisters, 28 of which are sufficient for a 14-day mission.

Lunar Module — The vehicle, consisting of the ascent and descent stages, which will transport two astronauts from the Command Module in lunar orbit, provide a base of operations on the lunar surface and return them to the Command Module; 19 feet tall by 29 feet wide, carried with legs folded in the spacecraft/LM adapter (SLA) during launch. (See Ascent Stage and Descent Stage.)

Lunar Module Pilot — One of the two men who will descend to the lunar surface. He is the primary expert on Lunar Module Subsystems. He occupies the right crew station in the Lunar Module and is primarily responsible for systems management.

Manned Space Flight Network — Worldwide network of 17 land stations (supplemented by 10 DOD Eastern or Western Test Range land stations, eight advanced range instrumentation aircraft, and eight ships), which supports Apollo manned flights with nearly continuous radar tracking, command signals, telemetry reception, and voice contact. MSFN, which includes the Mission Control Center in Houston, the Launch Control Center at Cape Kennedy, and a computing and communications center at Goddard Space Flight Center, is the responsibility of Goddard. Tracking stations are divided into three groupings: lunar mission support stations, equipped with 85-foot dual antennas; earth orbital and limited lunar mission support stations, equipped with S-band facilities; and the near-earth-orbital mission support stations, most of them modified Gemini network stations without S-band facilities.

Man-Rated — Adjective applied to spacecraft, test items such as a centrifuge, and test chambers, which have achieved the standards of reliability and safety considered acceptable for human occupancy or for use on a manned flight.

Memory — Portion of a computer; records and stores instructions and other data. Information is retrievable automatically or upon request.

Meteoroid — Solid particle of matter traveling in space at considerable speed. (See Micrometeoroid.)

Micrometeoroid — Solid particle of matter, less than a millimeter in size, traveling in space.

Mission Profile — Flight plan showing all pertinent scheduled events.

Multiplexer — Device for sharing of a circuit by two or more coincident signals; a device that collects data from many sources and arranges it for simultaneous transmission over a single network. That transmission is called multiplexing. The signals may be separated by time division, frequency division or phase division.

Nautical Mile — Distance of 6,076.1 feet, or about 1.15 statute miles.

Navigation Base — Rigid supporting structure for inertial measurement unit and telescope.

Nitrogen Tetroxide (N_2O_4) — Oxidizer used in the ascent and descent engines. The fuel used with a mixture of unsymmetrical dimethyl hydrazine and hydrazine.

Noise — Any unwanted sound or disturbance on a useful frequency band, which interferes with clear reception of radio or radar signals.

Non-destructive Readout — Readout of data stored in a computer memory; data is retained in memory.

Omnidirectional Antenna — Antenna having a nondirectional pattern in azimuth and a directional pattern in elevation.

Open Loop — Control system in which there is no self-correction as there is in a closed-loop system.

Orbit — Spacecraft's path around earth or the moon, beginning and ending at a fixed point in space and requiring only 360° of travel. (The point on earth where the orbit began will not be the same because, during the period of orbit, earth will have revolved in the same direction.)

Oxidizer — Substance that supplies the oxygen necessary for burning (normal burning on earth uses the free oxygen in the atmosphere).

Parallel Redundancy — Describes two components, methods, or systems working at the same time to accomplish the same task, although either could handle it alone.

Parameter — Characteristic element or constant factor or value; often, a limiting value or set of values.

Parking Orbit — Intermediate orbit around earth or the moon, where a spacecraft can await the proper moment for injection into a trajectory.

Pitch — Attitude movement of the Lunar Module, in which the Z-axis tips up or down, rotating around the Y-axis.

Portable Life Support System — Backpack containing oxygen, water circulation and cooling, air-conditioning, telemetry and communications equipment; worn during exploration of the lunar surface.

Premodulation Processor — Assimilation, integration, and distribution center for all forms of spacecraft data (telemetry, data storage, television, central timing and audio signals) and incoming voice and command signals. The processor mixes and switches the signals to the appropriate transmitter.

Pressure Garment Assembly — Space suit, including inner comfort layer, pressure layer, and outer restraint layer; boots; gloves; cloth earphone cap; and helmet.

Pseudorandom Noise — Signals, in the S-band frequency range transmitted from MSFN to the Lunar Module for ranging and tracking purposes.

Pulse-Code-Modulation Telemetry — Pulse modulation in which the signal is periodically sampled, and each sample is quantized and transmitted as digital code. Transmitted information is contained in the prime position of the pulse in relation to a known reference point. Pulse-code-modulation telemetry equipment in the Lunar Module combines signals from various sources into a single signal, which is sent to the premodulation processor.

Quick-Disconnect Fitting — Fitting designed for instant disconnection (umbilical cords, etc.).

Rate Gyro Assembly — Three rate gyros in the Lunar Module, which emit signals relative to indicating the rate of angular motion (attitude change rate) to the flight director attitude indicators and to automatic control equipment.

Real Time — As it happens. Term is usually applied to reporting of events as they happen or to computation of data as they are received, with nearly instantaneous readouts.

Redundancy — Alternative provision for accomplishing a task; as an adjective, redundant. (See Parallel Redundancy.)

Rendezvous — Meeting of spacecraft in orbit at a planned time and place.

Residual Water — Small amount of water in bottom of Lunar Module tanks not accessible for use.

Reticle Pattern — Pattern engraved on the crew optical alignment sight. Used in docking procedure.

Revolution — Circuit of earth or the moon beginning and ending at a fixed point on earch or the moon rather than a fixed point in space. Because the earth is revolving in the same direction, while the spacecraft is circling it, the point at which the revolution began has moved further ahead and the spacecraft must "catch up" with the reference point at the end of the revolution. A revolution is therefore more than 360° of travel and takes approximately 6 minutes longer than an orbit.

Roll — Rotation of the Lunar Module around its Z-axis.

S-Band — A 2100- to 2300–mHz band; carries voice, PCM telemetry, television, scientific data, coherent two-way Doppler, and tracking updata during all phases of the flight. In deep space, during the lunar mission, it is the primary voice link. Seventeen MSFN stations with unified S-band capability, are located around earth.

Signal-Conditioning Equipment — Devices that convert signals from sensors and transducers to proper format for transmission to MSFN.

Slant Range — Distance of the Lunar Module from the selected lunar landing site. Measured in a straight line from the landing radar antenna.

Slush Point — Temperature at which water-glycol starts to freeze.

Spacecraft — The Command, Lunar, and Service Modules, as distinguished from the Saturn launch vehicle.

Spacecraft Commander — Commander of the three-man Apollo crew; occupies the left couch in the Command Module and the left crew station in the Lunar Module. He is first in seniority and is trained in the skills of the Command Module Pilot and Lunar Module Pilot. He runs the mission from the standpoint of the crew, performs most of the engine burns, and is one of two men who descend to the lunar surface.

Spacecraft-Lunar Module Adapter — The 28-foot-high tapered cylinder between the Service Module and the launch vehicle instrument unit; it encloses the Lunar Module during launch and earth orbit. After translunar injection, a detonating fuse separates the Command and Service Module from the booster's third stage (S-IVB) and the Lunar Module. As the CSM turns around to dock with the Lunar Module, explosive charges and spring-loaded cables open the four hinged sections of the adapter like the petals of a flower. The CSM pulls the Lunar Module out of the adapter.

Stable Member — Major part of an all-inertial guidance system, composed of an assembly of gimbals that hold three accelerometers in a fixed position in relation to inertial space. The accelerometers are mounted perpendicular to each other to measure accelerations along the three reference axes. These accelerations can be fed to a computer to determine instantaneous velocity and position in space.

Staging (Stage Separation) — Separation of Lunar Module ascent and descent stages.

Station Keeping — Remaining in a particular, precise orbit with a constant velocity, usually at a given distance from a companion body.

Sublimation — Process utilizing space vacuum to transform ice to steam without first passing through liquid state. Lunar Module sublimators remove excess heat from water-glycol solution.

Telemetry — Technique of transforming sensed information into coded signals and transmitting it to a ground station, where it is decoded and fed into a computer for tabulation and readout. Telemetry measures the quantity or degree of such things as vehicle performance, medical information, temperature, pressure, radiation, velocity, heat rate, and angle of attack of the spacecraft.

Thrust — Push; the force developed by a rocket engine, measured by multiplying the propellant mass flow rate by the exhaust velocity relative to the vehicle, and expressed in pounds.

Thrust Chamber — Combustion chamber of a rocket engine; the place where fuel is burned in the presence of an oxidizer to produce high-velocity gases, which exit through the engine nozzle to produce thrust.

Thrust Vector — Direction of thrust. Thrust vector control is achieved by moving the gimbal-mounted descent engine so that the direction of thrust can be changed in relation to the Lunar Module center of gravity, producing a turning movement. (The Reaction Control Subsystem thrusters are mounted in sets and aimed in different directions, rather than on gimbals.)

Thruster — One of the 16 100-pound-thrust Reaction Control Subsystem engines used for attitude control of the Lunar Module. They are grouped in clusters of four. All use aerozine and nitrogen tetroxide.

Torquing Command — Command given to gyros to maintain vehicle attitude.

Tracking — Following a target by radar, optical sighting, or photography.

Trajectory — Flight path traced by vehicle under power or as a result of power.

Transceiver — Radio or radar transmitter and receiver combined into one unit, as is used in a transponder.

Transducer — Device that converts energy from one form to another; it is actuated by energy from one transmission system and supplies it to another system in a different form.

Transfer Tunnel — Passageway between Lunar and Command Modules when they are docked, for transfer of astronauts from one module to the other; reached by forward tunnel hatches in the Command Module and the overhead hatch in the Lunar Module.

Transponder — Radio or radar device triggered by a received signal of a certain frequency; transmits or returns the signals to the interrogator automatically. It is used in positive tracking and identification.

Tunnel Pressure — Pressure of the atmosphere (oxygen) in the tunnel connecting the Command and Lunar Modules (See Transfer Tunnel.)

Ullage — Volume above the surface of the liquid in a tank, partially a function of temperature. An ullage maneuver is a quick thrust of the vehicle made before firing the engines, to shift the propellant to the bottom of the tanks so that it will feed properly.

Umbilical — One of two electrical power cables connected between the Command and Lunar Modules before Lunar Module power is activated; hoses and electrical power cable between the pressure suit and vehicle and an oxygen line to the backpack.

Unsymmetrical Dimethylhydrazine — Component of Aerozine.

Updata Link — UHF/FM unified S-band receiver and decoding device (updata digital decoder); receives data from MSFN stations, decodes it, and routes it to the proper system.

Uplink Data, or Updata — Telemetry information from MSFN stations to spacecraft.

Vector — Magnitude of speed plus direction; short form of velocity vector, which is the speed of the vehicle's center of gravity at a certain point on the flight path and the angle between the local vertical and the direction of the speed. Vector control or vector steering is control of vehicle flight by tilting the descent engine to change thrust direction and produce a turning movement.

Velocity — Rate of motion (speed) in a given direction. (See Vector.)

VHF Multiplexer — Permits simultaneous transmission and receipt of VHF signals with a single antenna system.

Voice-Operated Relay — Transmit/receive circuitry, which is automatically switched to "transmit" by the sound of the astronaut's voice and returns to "receive" when the sound ceases.

Water-Glycol — Mixture of water and ethylene glycol; used to cool cabin atmosphere and space suits. It is, in turn, cooled by circulation through the sublimators.

X-Axis — Vehicle axis running up through the overhead hatch; associated with yaw maneuvers, in which the spacecraft rolls or spins around its X-axis.

Y-Axis — Lateral axis running through the spacecraft; associated with pitch maneuvers, in which the spacecraft turns or twists about its Y-axis.

Z-Axis — Fore-aft axis running through the spacecraft; associated with roll maneuvers, in which the spacecraft turns or twists about its Z-axis.

CONTRACTORS

EQUIPMENT SUPPLIED	CONTRACTOR
Abort Electronics Assembly Abort Guidance Section Abort Sensor Assembly	TRW, Inc. TRW Systems Group Redondo Beach, California
Absolute and Differential Pressure Transducers	Whittaker Corp. Instrument System Div. Chadsworth, California
Absolute Pressure Switch	Fairchild Hiller Corp. Stratos Div. Manhattan Beach, California
Actuator Bellows Assembly	Stainless Steel Products Burbank, California
Air Filter	Mectron Industries South El Monte, California
Ambient Helium Tanks	Sargent Industries, Inc. Airite Div. El Segundo, California
Ascent and Descent Batteries	Eagle-Picher Company Joplin, Missouri
Ascent – GOX Tanks Ascent Helium Storage Tanks	Sargent Industries Airite Div. El Segundo, California
Ascent Engine – Injector and Combustion Chamber	North American Rockwell Rocketdyne Div. Canoga Park, California
Ascent Engine – Skirt and Valves	Bell Aerosystems Co. Niagara Falls Blvd. Buffalo, New York
Ascent Propellant Tanks	Aerojet General Corp. Downey Plant Downey, California
Attitude and Translation Control Assembly	RCA Aerosystems Div. Burlington, Massachusetts
Attitude Control Assemblies	Honeywell, Inc. Aeronautical Div. Minneapolis, Minnesota
Bacteria Filter	American Air Filter St. Louis, Missouri
Bulkhead Feedthrough Connectors	ITT Cannon Electric Co. Los Angeles, California
Burst Disk	Parker Hannifin Corp. Systems Div. Los Angeles, California

EQUIPMENT SUPPLIED	CONTRACTOR
C-Band Transponder Antenna	Melpar Falls Church, Virginia
Cable Cutter Explosive Devices	Explosive Technology, Inc. Fairfield, California
Caution and Warning Electronic Assembly	Ambac Industries Arma Div. Garden City, New York
Caution and Warning Indicators	Penn Keystone Derby, Connecticut
Circuit Breakers	Mechanical Products Jackson, Michigan
Circuit Interrupter	ITT Cannon Electric Co. Phoenix, Arizona
Circular Connectors	The Deutsch Co. Electronic Component Div. Banning, California
CO_2 Sensor	Perkin-Elmer Electro-Optical Div. Norwalk, Connecticut
Coaxial Switches and Connectors	Quantatron, Inc. Santa Monica, California
Cold Plate Assemblies	AVCO Corp. Aerostructures Div. Nashville, Tennessee
Communication Subsystem	RCA Communications Div. Camden, New Jersey
Component Caution Indicators	Penn Keystone Derby, Connecticut
Control Electronic Section	Ambac Industries, Inc. Arma Division Garden City, New York
Coupling Disconnects	J. C. Carter Co. Costa Mesa, California
Coupling Test Points	Schulz Tool and Mfg. Co. San Gabriel, California
Data Storage Electronic Assembly	Leach Corp. Controls Div. Azusa, California
Descent Engine	TRW, Inc. TRW Systems Group Redondo Beach, California

EQUIPMENT SUPPLIED	CONTRACTOR
Descent Engine Control Assembly	RCA Aerosystems Div. Burlington, Massachusetts
Data Entry Display Assembly	TRW, Inc. TRW Systems Group Redondo Beach, California
Descent Propellant Tanks (LM4 and 5)	General Motors Corp. Allison Div. Indianapolis, Indiana
Descent Propellant Tanks (LM 6 and sub)	Sargent Industries Airite Div. El Segundo, California
Digital Uplink Assembly	AVCO Corp. Electronics Div. Cincinnati, Ohio
Diplexer	Ramtec Division Emerson Electric Company Celabasas, California
Disconnect	Fairchild Hiller Corp. Stratos Div. Manhattan Beach, California
Disconnect, Flight Half	Seaton-Wilson Co. Burbank, California
Discrete Transducers	Metals and Controls Div. Texas Instruments Inc. Attleboro, Massachusetts
Docking Lights	Eimac Div. Varian, Inc. San Carlos, California
Electrical Control Assembly	General Electric Co. Specialty Control Dept. Waynesboro, Virginia
Electroluminescent Lamps	General Electric Co. Miniature Lamp Div. Nella Park Cleveland, Ohio
End Detonator Cartridges	Space Ordnance Systems El Segundo, California
Environmental Control Subsystem	United Aircraft Corp. Hamilton Standard Div. Windsor Locks, Connecticut
Event Timer	Sylvania Electronics Needham Heights, Massachusetts
Explosive Nut and Bolt Assembly	Space Ordnance Systems El Segundo, California

EQUIPMENT SUPPLIED	CONTRACTOR
Explosive Valves	Pelmec Div. Quantic Industries San Carlos, California
Exterior Tracking Light	Dynamics Corp. of America Reeves Instrument Co. Div. Garden City, New York
Fire-In-The-Hole (FITH) Connector	ITT Cannon Electric Co. Phoenix, Arizona
Flag Indicators (Talkbacks)	Honeywell, Inc. Aeronautical Div. Minneapolis, Minnesota
Flex Lines	Avica Newport, Rhode Island
Flight Director Attitude Indicators	Lear Siegler Instrument Div. Grand Rapids, Michigan
Gimbal Angle Sequencing Transformation Assembly	Lear Siegler Instrument Div. Grand Rapids, Michigan
Gimbal Drive Actuators	The Garrett Corp. Airesearch Mfg. Co. Los Angeles, California
H_2O Bacteria Filter	Aircraft Porous Media Glen Cove, New York
Heat Exchanger (Discrete Engine)	Stewart Warner South Wind Div. Indianapolis, Indiana
Heater Assembly (RCS)	Cox and Co. New York City, New York
Helium Explosive Valves	Pelmec Corp. Quantic Industries San Carlos, California
Helium Filter	Aircraft Porous Media Glen Cove, New York
Helium Filter	Vacco Industries South El Monte, California
Helium Latch Valve	M. C. Manufacturing Co. Lake Orion, Michigan
Helium Pressure Valve	Parker Aircraft Corp. Los Angeles, California
Helium Quad Check Valve	Accessory Products Co. Whittier, California
Helium Relief Valve	Calmec Mfg. Corp. Los Angeles, California

EQUIPMENT SUPPLIED	CONTRACTOR
Helium Temperature Pressure Indicator	General Precision, Inc. Kearfott Systems Div. Little Falls, New Jersey
Helium Valve – Descent Regulator	Parker Aircraft Corp. Los Angeles, California
High Pressure O_2 Control Assembly	Parker Aircraft Corp. Los Angeles, California
Initiator	Space Ordnance Systems El Segundo, California
Interior Floodlight	Grimes Mfg. Co. Urbana, Ohio
Interrupter	ITT Cannon Electric Co. Phoenix, Arizona
Interstage Disconnect	Fairchild Hiller Corp. Stratos Division Manhattan Beach, California
Inverter	United Aircraft Corporation Hamilton Standard Div. Windsor Locks, Connecticut
Landing Gear Uplock Cutter Assembly	Space Ordnance Systems El Segundo, California
Landing Radar and Rendezvous Radar Subsystem	RCA Aerosystems Div. Burlington, Massachusetts
Latching Valve	Parker Aircraft Corp. Los Angeles, California
Lighting Control Assembly	Dynamics Corporation of America Reeves Instrument Div. Garden City, New York
Miniature Switch	Metals and Controls Div. Texas Instruments, Inc. Attleboro, Massachusetts
Mission Timer	Bulova Watch Company, Inc. Systems and Instrument Division Flushing, New York
Mission Timer	Sylvania Electronics Needham Heights, Massachusetts
Oxygen Fill Disconnect	Purolator, Inc. Aerospace Div. Newbury Park, California
Oxygen Hose	R. E. Darling Company Gaithesburg, Maryland

EQUIPMENT SUPPLIED	CONTRACTOR
Panel Overlay	Precise Engraving Company Garden City, New York
PLSS Condensate Collector Assembly	Lundy Electronics and Systems, Inc. Glen Head, New York
Portable Utility Light	Grimes Manufacturing Co. Urbana, Ohio
Potentiometer	Technology Instrument Corp. Newbury Park, California
Pressure Garment Assembly O_2 Connectors	Air Lock Corp. Melford, Connecticut
Pressure Relief Valve	Parker Aircraft Corp. Los Angeles, California
Program Reader Assembly	Fairchild Camera Space and Defense Systems Syosset, New York
Propellant Filter	Purolator Products, Inc. Aerospace Division Newbury Park, California
Propellant Filters	Aircraft Porous Media Glen Cove, New York
Propellant Level Detectors	Simmonds Precision Products Long Island City, New York
Propellant Quantity Gaging System	Trans–Sonics, Inc. Lexington, Maine
Propellant Quantity Indicator	General Precision, Inc. Kearfott Systems Division Little Falls, New Jersey
Propellant Quantity Measuring Device	Electro-Optical Systems Subsidiary of Xerox Corp. Pasadena, California
Propellant Solenoid Valve	Parker Aircraft Corp. Los Angeles, California
Propellant Tanks	Bell Aerosystems Co. Buffalo, New York
Pulse Code Modulation/Timing Electronic Assembly	Radiation, Inc. Palm Bay, Florida
Pushbutton Switch	Honeywell, Inc. Aeronautical Division Minneapolis, Minnesota
Pyro Battery	Electro-Storage Battery Raleigh, North Carolina

EQUIPMENT SUPPLIED	CONTRACTOR
Quad Check Valve	Parker Aircraft Corp. Los Angeles, California
Range/Altitude Indicator	Bendix Corp. Eclipse-Pioneer Div. Teterboro, New Jersey
RCS Explosive Cartridge	Space Ordnance Systems El Segundo, California
Reaction Control Subsystem	The Marquardt Corp. Van Nuys, California
Regulating Valve	Fairchild Stratos Western Branch Manhattan Beach, California
Relays	Filtors, Inc. Huntington, New York
Relief Valve	M. C. Manufacturing Co. Lake Orion, Michigan
Retractable Cable	Haveg Industries Supertemp Wire Div. Winooski, Vermont
RF Signal Sampling Sensor	Melpar, Inc. Falls Church, Virginia
Rotary Switch	Daven Div. Thomas A. Edison Industries West Orange, New Jersey
Rough Combustion Cutoff	Thiokol Chemical Corp. Reaction Motors Div. Denville, New Jersey
Self-Luminous Devices	Minnesota Mining and Mfg. Co. St. Paul, Minnesota
Sensor Probe	EDO Corp. College Point, New York
Signal Conditioning Electronics Assembly	Ambac Industries, Inc. Arma Division Garden City, New York
Signal Strength Meters	Honeywell, Inc. Aeronautical Div. Minneapolis, Minnesota
Solenoid Valve	Valcor Engineering Corp. Kenilworth, New Jersey
Steam Vent Divider	Stainless Steel Products Burbank, California
Suit Loop Switch	Parker Hannifin Corp. Systems Div. Los Angeles, California

EQUIPMENT SUPPLIED	CONTRACTOR
Supercritical Helium Tanks	The Garrett Corp. Airesearch Mfg. Co. Los Angeles, California
Surge Tank Disconnect	Seaton-Wilson Burbank, California
Synchros	General Precision, Inc. Kearfott Systems Division Little Falls, New Jersey
Target Assembly	Minnesota Mining and Mfg. Co. St. Paul, Minnesota
Thrust/Weight Indicator	Bendix Corp. Pioneer Central Division Davenport, Iowa
Time Delay	Lear Siegler Instrument Division Grand Rapids, Michigan
Toggle Switch	Metals and Controls Div. Texas Instruments, Inc. Attleboro, Massachusetts
Transducer	Hy-Cal Engineering Corp. Santa Fe Springs, California
Transistors	Metals and Controls Div. Texas Instruments, Inc. Attleboro, Massachusetts
TTCA Transducer	Bournes, Inc. Riverside, California
Universal Ball Joint	Stainless Steel Products Burbank, California
Waste Management System	Lundy Electronics and Systems Glen Head, New York
Waveguides	Electronic Specialty Co. Connecticut Division Thomaston, Connecticut
Window Panel Assembly	Corning Glass Works Corning, New York
Windows	Corning Glass Works Corning, New York
Wire	Haveg Industries, Inc. Supertemp Wire Division Winooski, Vermont
X (Cross) – Pointers	Honeywell, Inc. Aeronautical Div. Minneapolis, Minnesota

THE MOON

R155   NASA PHOTO

APOLLO 10 VIEW OF MOON – This photograph of the moon was taken after transearth insertion when the Apollo 10 spacecraft was high above the lunar equator near 27 degrees east longitude. North is about 20 degrees left of the top of the photograph. The terminator is near 5 degrees west longitude. Apollo Landing Site 3 is on the lighted side of the terminator in a dark area just north of the equator. Apollo Landing Site 2 is near the lower left margin of the Sea of Tranquility (Mare Tranquillitatis), which is the large, dark area near the center of the photograph.

It has been said that if it were not for the Moon, many facets of our life on Earth would differ from what they are today. For example, all moon songs would be about the satellites of Mars or Jupiter, since there would be none closer to write or sing about. Such catchy phrases as "moon glow" or "once in a blue moon" would be unknown. And imagine the plight of the lone coyote, limned against an evening desert sky, muzzle pointed toward the stars, but no moon to sing to!

More fundamental is the fascinating fact that if it were not for the Moon, life, that tenacious, complex, and irresistible force which swarms across the face of this planet in a million different forms, might not exist; and its highest product, Man, would not be poised upon the threshold of the greatest adventure since his ancestors' rejection of the sea, the Journey to the Moon!

The Moon is unique. This solar system, the sun and the nine known planets orbiting around it, contains at least three dozen satellites. Only Earth and its moon, however, have a size relationship to each other of four to one: Earth is 7,910 miles in diameter, the Moon, 2,165. All other planets are much larger in comparison with even their largest satellites or "moons." Because of this unusual closeness in planet and satellite diameters, the Earth-Moon system has been called essentially a "double planet," the only such system we know of.

It is this unusual similarity in size which first marks the earth and moon as unusual, but there are other observations. The moon moves around the earth in what is called an "elliptical" orbit; that is, it can be as near as 221,463 miles, or as far away as 252,710 miles. This raises another fascinating point. Of all the planets, Earth is unique in having a major satellite (and you have to call any satellite as big as the moon, major), orbiting about 30 diameters away. If one lists the characteristics of the other planets and their major moons, one usually finds an orbit distance of only 10 diameters away, compared to the primary planet.

A third highly illustrative observation of the Earth-Moon system can be made with regard to density. All density measurements—that is, how much material is contained within how much volume—are based on water as a standard. For instance, the density of a globe of water the size of the earth, with due allowance for certain physical effects, would be about equal to "1." The actual density of the earth—the average, that is—is really about 5.52. The earth, then, has a mean density about 5.52 times that of an equal volume of water. This figure, the mean density of a planet or a moon, can give valuable information as to chemical composition if used with care along with other physical characteristics.

The density of the Moon is 3.34. The Moon, therefore, is less dense than the earth, but denser than an equal sphere, 2,165 miles across, of water. It must be remembered, however, that these figures refer to the *average* density of the earth and moon. For the earth, we know the density increases with depth because the heavy elements, iron and nickel, have separated during the lifetime of our planet and have collected to form the core. According to Surveyor data, something similar, at least in principle, seems to have happened to the Moon. This last fact is extremely important in deciding between various theories concerning the origin of the Moon.

Thus, to an objective observer, the Earth-Moon system is quite an intriguing planet-satellite pair. The planet measures four times (approximately) the diameter of the satellite, which orbits roughly once a month at an average distance of 240,000 miles away. If the mass of the planet is taken as "1," then the mass of the moon which orbits it can be expressed as 1/81 that of the planet, or, it would take 81 moons to make an earth! This very unusual pair, spinning through space around a tiny yellow star which the inhabitants of Earth call "the sun" is, when compared with all the other planet-satellite combinations known in this solar system, unique. A logical question—considering that the inhabitants of Earth having set first foot upon their satellite—Where and when did it originate?

There are as many theories as to the origin of the Earth-Moon system as there are theorists, but, practically speaking, most fall somewhere within three main avenues of research: the moon originated from the earth; Earth and Moon formed jointly at the same time in close proximity to each other; or Earth at some time "captured" the Moon from some distant solar orbit.

To understand the formation of the earth and its moon, one really has to understand the origin of the entire solar system, as the one is obviously but a part of the other. Unfortunately, there are several internally consistent but mutually conflicting ideas on the latter, and it is in determining how this planet and the Moon originated that, strangely enough, will allow eventual resolution between them. Ergo, the often heard phrase: ". . . When we get to the Moon and can analyze a piece of it, firsthand, we'll not only find out where the Moon came from, but how this entire solar system originated." This is true. In science, it is called serendipity. *Any* information about the physical and chemical composition of the solar system can't help but provide additional information about its origin, or the origin of parts of it, because the system is precisely that,—a System! It is extremely fortunate that Earth has a Moon. If it did not, answers to these questions would have to wait until Man, provided he existed, could reach Mars or the asteroids. Analysis of the first truly uncontaminated extra-terrestrial material, the lunar samples,—in addition to shedding light on the age-old question of where the Moon comes from—will inevitably tell us something about where this planet and its satellite fit into the greater framework of the evolution of the solar system; and that makes the lunar landing but a Twentieth Century answer to a curiosity as old as Man himself: Where has he come from?

A star is being born. Gas and dust, swirled in great confusion, veil the fiery center of a solar system in formation. A series of lesser condensations, knots along the lanes of colliding atoms and solid grains, moves in spiral paths around the glowing proto-sun. Down there, coalescing out of the primeval interplanetary medium,—Earth. A flattened cloud, a rain of slushes: Ammonia, Methane, Water, and the like. Dust, debris and molecules, Hydrogen and Helium, a growing, spinning body . . . a planet. And nearby, according to prevailing thought, five billion years beyond this time, another object—a lesser eddy, also sweeping up material; lighter, smaller, destined to become a moon, a satellite forever of the other. Soon, in about a billion years, when the interplanetary skies are clear, and the nuclear fire of the central star burns steadily against the interstellar night, life will put in its appearance in the soupy, primeval mists of this young earth. The constant gravitational war, the conflicting fields of Moon and Earth down through the megayears, when the Moon is closer than she will be in 1969, raise huge tides, dwarfing those which Man will be familiar with; tides of "solid" earth, magma, molten rock, water, and even the primeval atmosphere of Hydrogen and Helium, race around the Earth. More megayears pass. Several fascinating results of this suspected close proximity of Earth and Moon are evident today. Earth's atmosphere is markedly different than it would have been if the Moon had not been near the earth at the time of its formation. As proof of this, we have only to look toward the brilliant "Morning" or "Evening Star," Venus, a sister planet about the size of Earth, but orbiting closer to the sun. Since Venus is about the same size, and our Mariner fly-bys have told us that its gravitational field is about the same as Earth's (close enough for this comparison), we would expect a comparable atmosphere, both in composition and in quantity. This is not the case! The atmosphere of Venus appears to be composed mostly of carbon dioxide at an atmospheric pressure at least one hundred times that found on Earth. An intriguing possible explanation of this discrepancy would be the earth and moon proximity during those early formative years. As the earth and its abnormally large, lone satellite whirled around each other across those megacenturies, the overlapping and constantly shifting gravitational fields inexorably stripped away the atmospheres of both new worlds, leaving a silent, rocky moon exposed to space, and a virgin planet, Earth, for volcanic action to mantle with new gasses,—nitrogen and carbon dioxide—in preparation across a billion years for Earth's first living cells; and later—much, much later . . . Man. According to our latest information, there is the distinct possibility that, without the presence of the Moon, there would have been no life on Earth.

Perhaps this is the "Why" of Man's fascination with the Moon, and our pilgrimage to her silent plains and wasted hills the repayment of a debt three billion years in age. But even in more recent history, from just two million years ago to Now (but a blink against the cosmic calendar), Man has worshipped Luna. In every language, the Moon . . . Selene . . . Luna . . . is personified as the Queen of the Night, Friend of Man, the Hunter, the Provider. If all the words in prose and verse and song were strung end to end,—and this is just a guess—they'd probably reach halfway to the Moon. Farmers around the world throughout recorded time have planted crops beneath her "correct" phase. Today, the tides she lifts, pale shadows of her former influence, govern fishing, sailing, and trade around this liquid planet. Science has even found a strange, but physical, connection between her waxing and waning face and our terrestrial rainfall. This is the lore and the lure of, as Shelly said it best, "That orbed maiden, with white fire laden, Whom mortals call the Moon . . . "

The waiting is almost over. The span of time remaining before Man's footprints mark forever the silent face of Luna can be measured now in hours. What will we find?

The Moon has not escaped unscathed her sisterhood with Earth. In 1969, she orbits our planet with an average speed of 2,287 miles per hour. At her present distance of 240,000 miles, this means a period of revolution of a little over 27 days; and, as you gaze at the Moon on clear evenings across that rift of space, you will notice the same features patterning the lighted aspect of her face. Luna

keeps one side facing Earth, the other side eternally averted . . . the Earthside and Farside of the Moon.

This came about when our satellite was much younger, probably because of its then much closer orbit to Earth. If the Moon raised the gigantic tides which it must have, on our planet, then Earth, with a mass eighty-one times that of the Moon, raised tides upon that body sufficient to cause a slowing of her spin until, today, five billion years after the presumed beginning of this double planet system, the Moon swings unceasingly around the Earth, her rotation locked in synchronization with her period of revolution.

Upon the side that Man could see across the centuries he noted, first with the naked eye and later with the telescope, that the face of Luna is divided into two distinctly separate areas: *Maria*, or the "Seas," and the land, *Terrae*. The Seas are dark, the highlands lighter, but the average reflectivity or albedo of the lunar surface is only about seven per cent. Full moonlight for earthmen is only about one millionth as bright as sunlight, but even this, biologists are beginning to suspect, is responsible for many intriguing aspects of Earth's ecology.

It was the invention of the astronomical telescope by Galileo Galilei, in 1609, that was to start a controversy which still rages 360 years after. Galileo described a series of " . . . ringed, walled plains, many of them perfectly circular, upon the Lunar Surface." The origin of the over 30,000 visible craters of the Moon was to become one of the most dramatic astronomical arguments of all time. Resolution must await manned exploration.

Time has stopped. This must be the first impression of the Apollo crew as they descend the ladder on the front leg of the LM and become the first men to stand upon the surface of the Moon. For a billion years, nothing has changed—the rounded rocks, the dead grey ground, the pitch black sky. The sun, a blinding, searing circle of white-hot liquid fire, will seem to hang motionless above the eastern horizon. Harsh contrasts, the glaring sunlit rocks, the long shadows of early morning on the Moon, will be filtered by the gold-plated visors of these first Earthmen to explore the surface of another planet. Above a stark horizon, prostrate before unfiltered rays of our parent star, the sun, a planet will slowly spin—a blue-green jewel, a swirl of atmosphere cloud and desert sands as Africa slowly rotates her vast expanse across the lighted side of Earth, Man's home, a quarter of a million miles away. Sixty-seven degrees above the sunlit western limb where lunar landscape meets the dark infinity of space, this cool, ocean-covered world is suspended by the Moon's eternal, synchronous rotation, a swirl of natural color visible above this vast, forbidding wilderness.

The Moon has paid her price for close association with the earth. Her small size and mass, resulting in her 1/6th gravitational pull upon her surface, as compared to Earth's, have doomed her to a life exposed to interplanetary space. Any atmosphere she might have had, collected during her formation, or released by subsequent activity from the interior, has vanished—lost forever because the molecules' and atoms' average speeds exceeded the small velocity needed to escape the clutches of her gravity.

This lack of atmosphere, this exposure to all radiation from the sun, the absence of wind, or waves, of rain or snow, is the single most important factor in determining conditions which Man will find upon the Moon. For unimaginable time, four billion circuits of our world around the sun, and more, the Moon and Space have met. The daily rain of particles, mostly small, sometimes large, debris left over from the dawning of the solar system, continually molds the surface of the Moon. At speeds never less than two miles per *second*, and sometimes as high as 100 miles per second, a constant flux, a meteoric rain, bombards the Moon, smashing, exploding, pulverizing the exposed surface of our only natural satellite. And what this "hailstorm" cannot do, our star's far-reaching atmosphere and radiation finish. The solar wind, that superhot, 100,000-degree extension of the sun in which our Earth and Luna orbit, sprays the surface of the Moon with atoms, nuclei, and ions. Its ultraviolet radiation, prevented from reaching the surface of Earth by oxygen miles above our heads, illuminates the constantly eroding rubble on the surface of Selene and, during periods of violent storms upon the sun, when tortured matter is ionized and flung at speeds near that of

light, itself, into the vast abyss of space, these particles and before them, X-rays, bathe the surface of the Moon with high intensity, ever-changing radiation, both particle and wave. Is it any wonder, then, that across those megayears, this treatment has produced the pulverized, crater-strewn, lifeless, dead grey ground on which Man will soon be standing?

There are 14,600,000 square miles of barren lunar isolation. Apollo 11, Mission G of Project Apollo, in fulfillment of the national goal set forth by then President of the United States, John F. Kennedy, will accomplish the lunar landing on 36 square feet of lunar surface. Where?

There were many constraints upon the choice of a lunar landing site. In, roughly, the order of their effect upon the final decision, these were: The site must be somewhere within 5 degrees north and 5 degrees south of the lunar equator, and it must not be more than 45 degrees east or west of the Moon's prime meridian. These two factors became obvious when it was decided to go to the Moon via what is called a "Free-Return-Trajectory," one that insures that if something happens to your motive power en route, gravity at the Moon will swing you around and back toward Earth.

The Moon is divided into two main terrain types, the highlands, mountainous, upland areas of the Moon, and *Maria* the misnamed "seas,"—vast, relatively level plains which appear more and more to be internal lava flows across the lunar surface, triggered perhaps by collisions of several small asteroids, 100 miles or less in diameter. Craters cover the lunar surface everywhere, but appear heavier on the highlands than on the seas. For these reasons, it was immediately apparent that Man's first Lunar Landing should take place on relatively level ground in an area where the local crater density was low and, if such a site could be found, near some intriguing surface feature; in other words, upon a *Mare*. The Ranger Program, on July 31, 1964, with the success of an 800-lb. unmanned spacecraft called Ranger 7 began a survey of several potential landing areas via live television as the spacecraft swept in toward a crash landing on the Moon at 6,000 miles per hour. These regions were, respectively, upon the western side with Ranger 7, on the eastern side with Ranger 8, and almost exactly in the center, in an ancient and scientifically interesting crater called Alphonsus, with the flight of Ranger 9. It was through the results of this successful Ranger series that we first became aware of the tremendous erosion caused by the constant exposure of the surface to interplanetary meteoric particles. This was possible due to the tremendous increase in resolution available from the Ranger television cameras over that attainable from Earth. The best

resolution acheived from Earth-based telescopes before the advent of lunar-bound spacecraft was about 1,000 feet. This meant that if a structure comparable to the U.S.S. Enterprise were to exist on the Earthside of the Moon, Man would never have been able to detect it through the constantly shifting atmosphere of Earth. It also meant that designers of the landing vehicle and trajectory planners could not know anything about the hazards or scientifically interesting features under 1,000 feet across. Ergo . . . Ranger.

R156

NASA PHOTO

APOLLO 10 VIEW OF MOON — An Apollo 10 northwestward oblique view of Triesnecker crater, centered near 3.6 degrees each longitude, and 4 degrees north latitude. This picture, taken from the Command and Service Modules, shows terrain features typical of northeastern Central Bay and the highlands along the northern margin of Central Bay. Beyond the highlands, the smooth floor of the Sea of Vapors extends almost to the horizon some 600 kilometers (375 statute miles) from the spacecraft. Triesnecker crater, approximately 27 kilometers (17 statute miles) in diameter, was 135 kilometers (85 statute miles) northwest of Apollo 10 when the picture was taken. The intersecting linear features to the right of the Triesnecker crater are the Triesnecker Rilles.

Grumman

Ranger photography made it possible to count craters on the lunar surface as small as five-tenths of a foot; but because of the nature of the missions, the area surveyed was obviously small and could only be observed for one "instant" in the history of the Moon. Something designed to simulate a manned landing, with cameras, was needed for the second phase of lunar exploration. This spacecraft, called Surveyor, would tell us if the Moon would bear the weight of a landing vehicle, or doom such a mission to disaster in jagged rocks or deep, smothering dust.

On June 2, 1966, with the Western Hemisphere of Earth in shadow, Surveyor I became the first terrestrial artifact to achieve a true soft landing on the surface of Selene. Its landing, in the extreme western edge of the previously described "Apollo Landing Zone" marked a fantastic "first" for Man and the United States. As Luna swept around the earth; from its vantage point on the Moon's Oceanus Procellarum, Surveyor became an extension of Man, watching the lonely and airless plain before its camera as the sun rose toward the zenith, then descended in the West, finally to set, plunging the spacecraft into the numbing two-week night upon the Moon. It saw, and sent to its creators, the panorama of a surface turned and battered, of blocks and rubble, and of rounded mountains, the eroded remains of an ancient crater rim beyond the east and north horizons. Surveyor photographed the stars from the surface of the Moon and, after sunset, watched the breathtaking view of our sun's own atmosphere, its corona, visible to earthbound astronomers only in eclipse, sink slowly beneath the western limb, the line separating Moon from starry space. And finally in the early hours of a night that would last fourteen days, sixteen hours, and fifty-one minutes, Surveyor sent to Man across the quarter million miles separating him from his explorer, his first view of earth-light from the Moon.

Surveyor I proved that, at least in the Moon's Ocean of Storms, Man could land and walk upon our satellite. There was no deep dust. Rocks and craters there were, but these should not prove especially troublesome to a manned landing, since the final landing site would be determined by the pilot of the LM, and the spacecraft was being designed with some hover capability. The Moon seemed very near in the early morning hours of June 2, 1966.

Concurrent with the Surveyor unmanned soft landings, another program was to become operational, the unmanned Lunar Orbiters. These spacecraft were designed to fill the gaps between the Ranger series and Surveyor, by providing extensive orbital reconnaissance of all proposed Apollo landing sites, as well as orbital photography of Surveyor landing sites, actual and potential. In this way, information obtained by the Surveyor series could be extrapolated to include the vast areas Surveyor couldn't see or sample.

The final Lunar Orbiter, Number Five in the series, completed picture readout of its last lunar photograph August 27, 1967. The Orbiters, as contrasted to Ranger and even Surveyor, did not take pictures via a direct television process, but, through an ingenious 150-lb. orbiting photographic camera/laboratory system. All photographs, taken directly on special high resolution film, were developed as the spacecraft orbited the Moon, then read back to Earth after a conversion of the film information to electronic signals. On Earth, these signals were used to run equipment which did the reverse, converting the electronic analogs back into light and dark areas on film. The same technique was first tried around the Moon when the U.S.S.R. sent its Lunik III looping around our only satellite, but the state-of-the-art used in Orbiter was vastly improved over the early Russian attempt.

In all, five Lunar Orbiter spacecraft were sent into various orbits around the Moon. The first three were used expressly to scout the potential Apollo landing sites along the lunar equator. Because of the eminent success of these three, the remaining two were used to go after areas of scientific interest not directly concerned with the immediate goal of an Apollo landing before 1970. These spacecraft were injected into very highly inclined orbits, lunar polar orbits in fact, and made Man's photography of the entire Moon, including those areas never seen from Earth, ninety-nine per cent complete.

It was the Orbiter series that was also to make a very important discovery, from both engineering and scientific points of view, concerning the lunar interior. As the spacecraft, especially Orbiter 5, orbited the Moon, tracking was accomplished via the established ground stations of NASA's Deep Space Instrumentation Facility. With this data from all five spacecraft, it was hoped, a much better idea would be obtained concerning the nature of the lunar gravitational pull and how it behaved, in fine detail, as this would reveal information about the distribution of the Moon's mass in its interior.

From tracking data obtained during the mission of Lunar Orbiter 5, two investigators, Paul Muller and William Sjogren, of Cal Tech's Jet Propulsion Laboratories, revealed a series of denser parts of the Moon, "Mass Concentrations" or Mascons, discovered as the spacecraft, in polar orbit, traveled over five of the circular *Maria* visible on the Earthside. The irregular *Maria*, on the other hand, had much less effect on the spacecraft orbit.

While these results are fascinating from an astrophysical point of view and shed new information on theories of *Maria* formation, they also presented Apollo with a problem. As the Lunar Module is making its final descent from 50,000 feet to the lunar surface, it must fly over the Mascon areas, thereby falling prey to their perturbing effect on its trajectory. Apollo 8, orbiting men for the first time around a celestial body other than the earth, demonstrated the effect of these Mascon areas on a manned spacecraft in lunar orbit. Apollo 10, capable of accomplishing a lunar landing, but not equipped to make one, provided exact gravitational analogs by orbiting at similar heights and ground tracks as the mission to follow—Man's First Lunar Landing.

At the completion of the Orbiter program, the initial Apollo landing sites had been narrowed to five, scattered from east to west across the "Apollo Belt." The selection process at this stage of the game had excellent, high resolution photography with which to map the local terrain, analyze crater density, the heights and depths of terrain features, and areas of scientific and geological interest for the initial landing. Because of the long, very shallow descent profile followed by the LM as it rifles in toward the target area, special care had to be taken to insure the proper approach terrain along the ground track some sixty miles east of the actual landing site. This was due to the necessity of certain radar returns, which give the LM its velocity of descent as well as its height above the surface. Mountains, ridges, or large craters would cause anomalous readings, thereby "misinforming" the guidance computer of the spacecraft's actual position during those critical minutes. Many otherwise acceptable landing sites had to be discarded because this constraint made it impossible to "get there, from here, this way."

R157

NASA PHOTO

APOLLO 10 VIEW OF MOON — This near vertical photograph taken from the Apollo 10 Command and Service Modules shows features typical of the Sea of Tranquility near Apollo Landing Site 2. The proposed landing area for Apollo 11 (Lunar Landing Site 2) is a relatively smooth maria area in the upper right quadrant of the photographed area. Apollo 10 traveled from the bottom to the top of the picture. The prominent linear feature on the left is Hypatia Rille (called "U. S. 1" by the Apollo 10 crew). The prominent crater centered in Hypatia Rille at top left is Moltke AC (code name "Chuck Hole"). Moltke, the prominent crater to the right of Hypatia Rille, is centered near 24.2 degrees east longitude, and 0.6 degrees north latitude.

Even at the landing site, an area, not a point, had to be chosen. The actual spot at which the first men set foot on the Moon will be somewhere within an ellipse 4.5 miles long by 3.2 miles wide. This wide spread in the actual landing area is due to unavoidable errors in the instrumentation and in our imperfect knowledge of the effect of the Moon

on the landing descent trajectory. It is an excellent maxim to remember, that the more we land on the Moon, the better we'll become.

At this writing, the most probable point on the lunar surface where men will first stand is somewhere within that ellipse, 4.5 miles long by 3.2 miles wide, centered at 23 degrees 37 minutes east

R158

NASA PHOTO

APOLLO 10 VIEW OF MOON — An Apollo 10 westward view across Apollo Landing Site 3 in the Central Bay. Apollo Landing Site 3 is in the middle distance at the left margin of the pronounced ridge in the left half of the photograph. Bruce, the prominent crater near the bottom of the scene, is about six kilometers (3.7 statute miles) in diameter. Topographic features on the surface of the Central Bay are accentuated by the low sun angle. Sun angles range from near six degrees at the bottom of the photograph to less than one degree at the top of the photograph.

Grumman

longitude, zero degrees 45 minutes north latitude. It lies in the southwest corner of *Mare Tranquillitatus,* the Sea of Tranquility. It is an old *Mare*, a lava plain formed about four billion years ago, and pounded and overturned to a depth of almost thirty feet. It will be flat, not too cratered, old, and weathered by the megacenturies' exposure to a vacuum more perfect than any possible beneath our atmospheric sea. The landscape will be grey, dead, lifeless. Across the horizon to the north, not 40 miles away, lies a crater in the wilderness, a fresh new scar upon the ageless face of Luna. Here, at 6,000 miles per hour, a part of history flashed silently into oblivion. In the sunlight, a few twisted metal fragments glint in testimony to the pioneer that was Ranger 8.

To the northwest, sixteen miles across another part of this flat and featureless horizon, is another monument to the genius that is Man. This artifact stands tall and proud upon the silent plain, its panels flashing in the sun, its golden valves and painted blocks a colorful display against the ashen surface. This is Surveyor V, and from it, gently resting on the crater rim, a box, lustrous and golden by the yellow glare of Sol. With this instrument Man made his first direct chemical analysis upon another world and found the Moon to be made, not of cheese, but of basalt, of once-molten lava, pulverized by time and exposure to the elements which rain upon this unprotected plain across eternity. Not too far from here, two men will stand, quite soon now, beneath a canopy of sun and earth and stars, emissaries to this hostile alien world from the cool green hills of earth, a quarter of a million miles across the sky.

The reasons for Man's Journey to the Moon are elegant in their simplicity. Man is going to the Moon because he can. To do otherwise would be a refutation of his past 2,000,000 years upon this planet. The history of Man has been the history of Man's life-long struggle for better understanding and control of his environment, of its utilization for his economic progress through better information and advancing technology. Man has now attained the capability to live anywhere upon the face of this third planet of the sun—this Earth—that he may choose to live. His vessels can descend into the deepest rifts of any ocean. His stations, powered by the fissioning of atoms, stand amid the ice of both the polar caps. Surveillance satellites, powered by the radiation of the sun, sweep around the planet, photographing weather in detail. As evidenced by history, the acquisition of knowledge and the control of energy have been the two most important factors governing the dominance of Man upon the earth. When viewed in this perspective, the Journey to the Moon is only the beginning of a future in which Earth's satellite and all the solar system will have been acquired for the betterment of Man, both on and off his home.

It has been said that the Moon is the most valuable piece of real estate man has yet acquired, and that exploring it will answer many questions: the origin of Earth and Moon, the solar system, even Man,—questions that today are only dimly phrased from the confinement of this sphere. If only for its intrinsic value as a chronicle, the Moon is priceless. Time is a terrestrial destroyer. The only universal constancy—Earth has guarded jealously the secrets of the past. Forces deep inside the earth, convection currents within the plastic mantle of the world, constantly give rise to mountains while dooming others to extinction. Upon the surface, the constant interplay of air and sea, the never ending cycle of the oceans, from waves to evaporated molecules, to droplets, and thence to rain and hail and snow, sluice away the record of the ages, wearing down the rooftops of the earth and carrying the silt back to the ocean floor. Life, in the form of plants and microbes, animals, and even Man, destroys each yesterday forever in pursuit of each tomorrow. The earth is a seething cauldron of hurricanes and tidal waves, geologic change, earthquakes and erosion, wind and pounding surf against the shores of every continent and island. And what these forces cannot grind away to dust, the atmosphere will slowly oxidize and change through combination. Mountains rust away, acids eat at geologic strata, and the layer of sediment buries everything as the planet, youthful still, evolves across the aeons.

In the midst of this inconstancy, Man has tried to piece together fragments of his heritage, to find a clue to the origin of worlds and suns, and of himself. The wonder is that he has done so well, in spite of the annihilation wrought by Mother Earth on artifacts of planetary history. Even a matter as basic as the abundance of the cosmic elements cannot be learned from study here. Through the process of formation, Earth effectively concealed this record: the heavy metals sank, because of gravity, to form the iron core of this, the sun's third planet. There they stay, locked, perhaps forever, beyond the probe of instruments. This makes any counting of the atoms in the crust hopeless for determination of the original distribution of the elements that formed our planet and the system, as a whole.

It is on the silent face of Luna, amid the desolation that has never known the fall of rain or the pounding of the sea that these clues to a yesterday removed forever from this earth will be discovered. There, too, change has left its mark: volcanic and external, the flow of lava from within and the explosive blast of meteoric impact from without.

But this change is of a different quantity and kind than that which has shaped and shaped again the surface of our world. Geology upon the Moon will tell a fascinating story of its formation and the forces which have made it what it is today. The surface rubble, from its depth and consistency, will disclose the frequency of objects orbiting the sun that had rained upon that surface, and the radiation character of every excavated inch will reveal the flux of cosmic rays and solar storms across the megacenturies our satellite has swept around the earth, waiting for the day when Intelligence would bridge the gap between these worlds, the planet and its satellite, and ponder its discoveries.

For the next several centuries, Man will be learning from the Moon. There is even speculation that beneath her bone-dry plains—ironically, the "seas" of Galileo—Man will find the molecules that in the liquid seas of Earth, began the spark of life. Perhaps, upon the Moon, suspended since the time of their formation four billion years ago, these complex organic compounds have remained, waiting for discovery by descendants from across 40,000,000 centuries of time and a quarter of a million miles of space. The discovery of proto-life, the step just before the creation of that first, simple, self-replicating cell, would undoubtedly rank as the greatest scientific discovery of this century, just short of the discovery of extra-terrestrial life itself. In all the solar system, after

due consideration for the environments of all the other planets and their satellites, this discovery of pre-organic molecules preserved against the ravages of time, is only possible amid the isolation and the silence of the Moon.

But even if it weren't for her intrinsic value, as the Rosetta Stone that will unlock the mysteries of Time, the Moon would still become a cherished gift to Man, as the foundation for exploration of the Universe.

Astronomy and astrophysics cannot help but benefit from the establishment of an observatory on the Moon. Earthbound astronomers for centuries have seen " . . . but through a glass, darkly," their view of space only a distorted glimpse through the thick, shimmering, absorbent veil of Earth's dynamic atmosphere. The limitations imposed by observation through this interfering gaseous mixture are almost too numerous to mention. It, first of all, absorbs most of the energy emitted by objects in the Universe: stars, galaxies, and quasars—just to name a few. What energy does finally filter down to the surface of the earth is then distorted through refraction imposed upon this radiation by our "transparent" air. This effect, the cause of "twinkling" of the stars, limits the fineness of detail that can be seen or photographed, be it of an object a million light-years distant, or right next door, such as the Moon.

Sometimes, on nights when skies are cloudy, the only radiation from the heavens reaching Earth is the delicate whisperings of radio emissions from the stars. The observatory on the Moon, with an array of telescopes covering the spectrum,—optical, X-ray, infrared and radio—will advance Man's knowledge of the Universe at least a thousand years. Anchored deep inside the lunar crust, these instruments will have the capability of seeing all of space in a period of time (27 days) equal to the rotation of Selene, if placed at the equator. Their gaze will span a distance back across the light years to the Dawn of Time itself, when the Universe was born. Time exposure photographs, measured not in hours, but in days, will image for the eyes of Man the details of the birth of stars and galaxies impossible to view from Earth observatories.

From that airless lunar plain, totally exposed to all the silent radiations that flash across the interstellar night, Man will first attempt to photograph the planets of the nearby stars. With telescopes at least six times as large as any ever built before, the light of other suns will be collected. By careful screening of their glare, a search in their vicinity for tiny, planetary points of light will then commence. It is thought that almost all stars form planetary systems as the natural function of their origin. Detection of these worlds in actuality,—a feat impossible except from space—would lead to great advances in our understanding of the Solar Family. It would also provide us concrete targets for our probes across the dark light-years, and may even lead to detection of a second "Earth," somewhere, orbiting an alien sun.

Sometime within this century, as Man explores and develops the resources of the Moon, there will be established upon the Moon's Farside, an additional observatory. This will be done in desperation as an increasingly technical society on Earth broadcasts its presence to the stars. The electromagnetic interference from television, radio, aircraft, surface transportation, and the like, is now, in 1969, a roar. In future years, this din will rise at an increasing rate until it drowns all radio reception from the Universe beneath a sea of electronic noise.

Radio astronomers, searching desperately for a solution, will plan and implement a research station on the Moon's Farside, forever out of sight of Earth; for even across the intervening quarter of a million miles, terrestrial electronic interference with delicate detection of intersteller radio emissions cannot be tolerated. A radio observatory on the Farside of Selene, with at least a thousand miles of solid rock screening it from Earth's electromagnetic clamor, will scan the heavens unhindered, mapping the natural song of hydrogen, hydroxyl radicals, water, ammonia, and a dozen other compounds, deciphering the death throes of a star or the birth-pangs of the Universe, itself. And from the isolated silence of the Moon's Farside, this observatory could become the first receiver of a signal *not* of natural origin, but one sent by Intelligence in search of a reply across the sea of stars which forms the Milky Way.

If utilization of the Moon is important, astronomically, to science, then so will be discoveries that are to come from basic research conducted in her physically unique environment. The basic properties of matter, the structure of the nucleus, even the laws governing the existence of Space and Time themselves can be discovered through the proper questions asked upon the surface of our satellite. When laboratories are constructed in the space environment upon Selene, physicists will have available for interaction with selected target materials, the most powerful radiations known in all the Universe, primary cosmic rays. These particles, ejected from the star-rending explosion of a sun, are accelerated to velocities near that of light. By studying their impact, science will come ever closer to an understanding of the forces which have shaped Creation, under conditions unattainable on Earth.

In laboratories on the Moon, research will also center on the effect of space on matter. Here, for mile after mile, is a more perfect vacuum than can ever be economically attained beneath our Terran atmospheric sea. On Earth, each cubic inch of "nothing" must be paid for dearly through the use of complex seals and pumps and power. On the Moon, all the vacuum Man will ever use is there, as free as air on Earth.

Scientific exploration of the Moon implies survival on the Moon; a base, with a variety of personnel, supplies, and means of manufacture and construction amid the Moon's environment. The philosophy whereby one transports everything,—men, equipment, food and water,—to the lunar surface must someday give way to means of utilizing those available resources on the Moon itself to sustain tomorrow's colonists upon her airless plains.

Apollo will determine, on a *Mare*, what the Moon is made of; but this determination will be essentially confined to scattered points on the top-most layer of a planet having an area almost as great as North and South America combined. It is obvious that a detailed inventory of the Moon's resources must await much more extensive exploration than the initial landings.

In areas of volcanism,—and from Orbiter photography, these are almost certain to exist,—those environmental factors favoring the establishment of a self-sustaining Lunar Base are most likely to be found. There will be lava tubes, allowing structures to be inflated underground, away from the 500 degree extremes of temperature encountered on the surface, as well as the eventuality of meteoric impact. Given enough energy (either the sun which shines full-force upon the Moon, or nuclear sources), Man can extract those vital elements of life, oxygen and water, from the surface of Selene. According to Surveyors 5 and 6, over half the atoms present in the surface were revealed as oxygen. Water was not detected in these historic findings, but should exist, especially if volcanism plays a part. In certain geologic forms, water should form as much as ten per cent by weight, allowing economical extraction through the use of energy derived from reactors or the sun.

This will be the key to the development of all the Moon. If mining and, through the use of energy, the subsequent extraction and refinement of essential air and water is possible upon the Moon, then, before the year 2,000 the Moon will be acquired for the betterment of Man. Colonies of scientists and engineers, of businessmen and manufacturers, of pioneering families, will transform Selene into a vital necessity to the people of the earth. Industries impractical forever here on Earth will flourish on the Moon where instant vacuum, temperature extremes, and 1/6 gravity lead to technological discovery. In the electronics field, alone, these conditions will inevitably lead to the growth of a whole new era of ultra-pure, solid-state components. With abundant energy, sophisticated treatment of the Moon's raw materials may lead to entirely new metals, plastics, glass, to say nothing of the manufactured products built around these new discoveries.

The once barren lunar surface will be transformed; clusters of domes rising from the airless ground in places the names of which have rung through history: Copernicus, the Apennines, the northern shores of Mare Imbrium. Monorails, gleaming in the two-week sun, will flash silently along, ferrying passengers and freight from the Observatory to manufacturing complexes, from tourist centers to the spaceport, across the wilderness, the Mare, and Terrae, from pole to pole, and even to Farside. Whole cities, peopled by the citizens of Luna, will grow beneath the surface of Selene. Before the turning of the century, amid the brisk development of this entire double-planet system, children, many never to set foot upon Man's green oasis, Earth, will be born upon the Moon. They and their descendants will make the Moon and other planets of this solar system a vital part of Earth's economy, feeding, housing, and supplying the people of the sun's third planet with many of their needs.

Realization of the Moon's potential, its resources and environment, must include a study of its effect on spaceflight as a whole. Escaping from the Earth, against its gravity, is an expensive operation which will deter efficient use of space until a way is found to overcome the inefficiencies of present rocket systems.

Todays massive vehicles, the Saturn 5's, the Titans and those similar, will soon—must soon give way to vehicles designed to be reusable; vehicles that will leave the surface of the earth, attain orbit just outside the atmosphere, and then return, landing like an aircraft of today. These ferry systems, carrying passengers and cargo into low orbit of the earth, will rendezvous with cities in the sky, space-stations orbiting the planet at altitudes around 500 miles. There, equipment, personnel, and vital luxuries unobtainable on the Moon, will be transferred aboard a Lunar spacecraft, a vehicle designed in the pattern of the LM, never to make a planetfall upon a world with atmosphere. Using fuel manufactured on the Moon, this transport system will follow its return trajectory to Luna and in three days will land its cargo upon the airless surface of our satellite. Man will learn, even if he does not realize it now, how unfortunate it is that Earth has a satellite located as it is. Rich beyond belief in minerals, information, and environment, it will someday become more economical to transport products, men, and even food from the surface of the Moon, a quarter of a million miles away, than it is or will be then to travel from the earth into an orbit of the earth. This will be possible because of the Moon's small mass, its lack of atmosphere, and its position.

An object, molecule, or spacecraft needs almost seven miles per second to escape from Earth completely. To accomplish this same feat from the surface of the Moon requires only one point five. This means that, pound for pound, it is some twenty-five times easier to escape Selene than Earth. Because of this, and the fact that Luna is eternally exposed to space, our Moon will undoubtedly become a launching platform unequalled in the solar system.

It is on the Moon that a spacecraft-launching system impractical on any other planet can be used. A catapult, a gleaming steel and concrete bow, will stretch across the miles of lonely lunar wilderness. Powered by the sun or man-made tapping of the atom, this narrow silver arrow will become Man's means of travel to the limits of this system and beyond. Visible from Earth by even simple telescopes, this shining strip of steel will hurl, without resistance from the airless lunar surface, capsules containing medicines, materials, and products developed on the Moon, to Earth. Its power will accelerate and fling into the emptiness, manned craft to journey to the planets, carrying exploration of these captives of the sun to its logical conclusion: the acquisition of these places, too, for Man. Finally, from the timeless lunar landscape, with the almost unlimited capacity of this device anchored to the Moon, the instruments of Man will rise around her curving, silent arc, and fall across the Night which separates the stars. And someday, men will follow.

THE AUTHOR

Richard Hoagland is a former staff lecturer and Curator of Astronomy and Space Science at the Springfield Museum of Science in Massachusetts. He subsequently was Assistant Director of the Gengars Science Center and Planetarium at Children's Museum, Hartford, Conn., and devised several major programs to modernize planetariums in the U. S. His innovations include techniques described as "a major breakthrough in the field of planetarium programming and simulation" in the journal Sky and Telescope. A writer and lecturer, Mr. Hoagland is a consultant on astronomy and space science to museums, planetaria, and the aerospace and broadcasting industries.

LUNAR MODULE DERIVATIVES FOR FUTURE SPACE MISSIONS

The assets of the Apollo Program collectively represent a major resource for future United States space missions. These assets include test, production, tracking, and launch facilities; trained astronauts; spacecraft; launch vehicles; and the experienced government/industry/university team.

In particular, the Lunar Module's versatility, ability to fly manned or unmanned, propulsive capability, and payload volume envelope offer attractive options to achieve significant space objectives in the future. These include increases in scientific knowledge, benefits to Man, space technology, and mission time.

A variety of LM-derived vehicles for use in earth or lunar orbit, and on the lunar surface, are described in this section. Missions of 2-month duration in earth orbit (longer with revisits) and at least 2-week duration on the moon for a variety of important space objectives are possible utilizing the LM. These missions would apply the experience gained from the early LM mission, to reduce costs while greatly increasing mission and scientific objectives. Additionally, these missions could serve as the springboard for significant future space ventures.

APOLLO LUNAR MODULE

This—the basic Apollo Lunar Module, or LM—is the two-stage spacecraft from which all the concepts shown are derived. For Project Apollo, the LM mission is essentially this:

Ferry two astronauts to the surface of the moon from the Command and Service Module (CSM) parked in lunar orbit.

Sustain the lives of the men during their lunar stay

Exploration and gathering scientific data and samples

Return the astronauts safely to the CSM through an ascent and rendezvous maneuver.

The descent stage carries the equipment and expendables used during the descent to the moon and lunar exploration; the equipment includes the engine, propellant, the landing gear, and the Apollo Lunar Scientific Experiment Package. The ascent stage contains the crew's life support equipment, expendables, and storage provisions for return of scientific samples; this stage contains most of the spacecraft's other operating subsystems including the engine and propellants used to return the astronauts to the CSM. As designed for the initial Project Apollo missions, LM was capable of sustaining the astronauts for up to 48 hours away from the CSM, carrying a 300-pound scientific payload to the moon, and transporting a 100-pound payload on its return trip to the CSM.

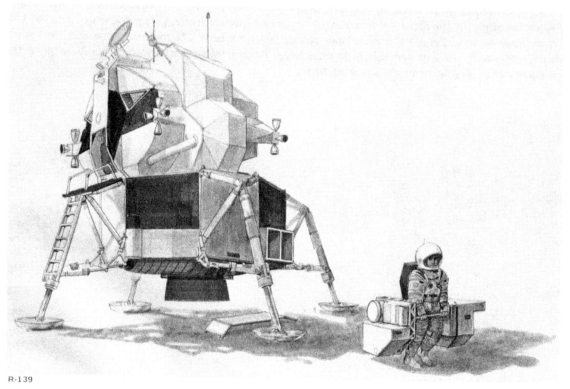

R-139

GRUMMAN

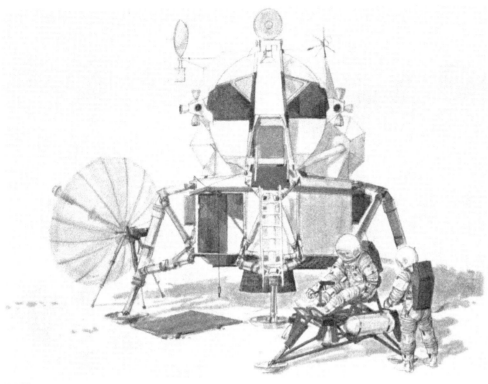

R-140

EXTENDED LM

For single-launch missions, experience gained from the initial Apollo lunar landing, combined with some LM modifications, permits an increase in LM payload. These modifications extend astronaut time on the moon to 72 hours while providing the astronauts with as much as 1000 pounds of scientific equipment, including Lunar Roving Vehicles (LRV's) or possibly Lunar Flying Vehicles (LFV's).

LUNAR RECONNAISSANCE MODULE

The Lunar Reconnaissance Module (LRM) is an orbiting version of the LM equipped with extensive photo-mapping, geochemical, and electromagnetic surveying equipment. Docked to the CSM, it is inserted into a lunar polar orbit from which, for the first time, the surface, subsurface, and near-lunar environment is surveyed.

The LM descent stage is used for initial Lunar Orbit Insertion (LOI), after which it is jettisoned. The ascent stage, stripped of its Propulsion and Reaction Control Subsystems, contains all the sensors for the reconnaissance mission.

The LRM carries its own equipment thermal control system and electrical power supply system, in addition to navigation and control equipment for the descent engine docked burn.

The LRM completely surveys the moon during 14 days in lunar orbit. Normal equipment servicing, such as film reloading, is accomplished in a shirtsleeve environment.

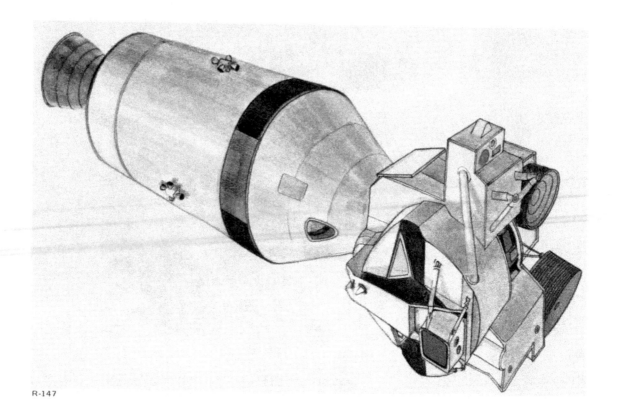

R-147

LM LABORATORY

Heavily equipped with experiment instrumentation, capable of operation in earth or lunar orbit, and provisioned to sustain two astronauts for 45 days in space, the LM Laboratory is an exciting science-oriented offshoot of the LM. Its experiment sensors include radiometers, spectrometers, a stellar camera, a terrain camera, multispectral cameras, X-ray sensors, a day-night camera, and an IR imager. Experiment categories in which these devices are used include meteorology, astronomy, earth resources, lunar survey and mapping, bioscience, and engineering technology.

The laboratory is compatible for docking with the CSM. It can be launched by the Uprated Saturn I or Saturn V vehicle.

R-141

LM TAXI

Outwardly identical with the Apollo LM, the LM Taxi differs from its parent spacecraft by a few subtle modifications to accommodate the slightly different nature of its mission. The LM Taxi ferries two astronauts to the moon on the second leg of a dual launch, following the successful unmanned landing of a LM Truck/Shelter on the first leg.

The Taxi carries the same life support provisions as the Apollo LM. The astronauts will shut down and store the Taxi after landing, and transfer quarters to the Shelter vehicle for the 14-day stay. During this period, the status of mission-critical Taxi hardware is monitored on earth via MSFN.

During quiescent storage, the propellants, engines, and much of the Taxi equipment must be thermally controlled, to permit rapid abort. The Apollo LM thermal control system has been modified to include a hatch cover, window shades, and isotope heat sources. Additional instrumentation has been added to enable complete evaluation of Taxi status at all times. A Radioisotope Thermoelectric Generator (RTG) supplements the battery power supply capability. All subsystems are qualified for the extended life requirements of the Taxi.

R-142

LM TRUCK

The LM Truck is an unmanned lunar lander that transports cargo in the volume otherwise occupied by the LM ascent stage. Components from the removed ascent stage that are vital to the Truck mission are relocated in a central docking structure attached to the existing interstage fittings on an unmodified descent stage.

The docking structure enables CSM transposition and docking with the Truck in lunar orbit and extraction of the Truck from the Spacecraft LM Adapter (SLA). Thereupon, an astronaut from the CSM can reach into the structure through the docking tunnel and, using a keyboard stored in the structure, update the Truck's Guidance, Navigation, and Control Subsystem so that the spacecraft's trajectory to landing is held within acceptable tolerances.

A typical mission payload, shown in the artist's rendering, might comprise a Lunar Roving Vehicle for surface transportation, resupply modules for supporting two men on the lunar surface for as long as 14 days, and a 5,300-pound, 900-cubic-foot scientific cargo.

GRUMMAN

R-145

LUNAR PAYLOAD MODULE

Unlike the LM Truck, the Lunar Payload Module retains the basic LM ascent stage, but is stripped of the ascent propulsion system and components unnecessary for lunar landing. In this configuration, the existing ascent and descent stage structures can accommodate a 7,300-pound payload within a useful volume of approximately 800 cubic feet.

Otherwise, its mission is similar to the other shelter vehicles launched at the start of a dual-spacecraft mission. The payload module, landed unmanned, replenishes shelter and astronaut supplies.

LM SHELTER

The Shelter is an Apollo LM minus its ascent propulsion system and modified to: (1) make an unmanned landing on the moon, (2) remain quiescent for as long as 60 days, and (3) support two men for 14 days. Successful launch and landing of a Shelter would be followed by a manned Taxi in a dual-launch mission. Shelter payload could consist of expendables, mobility aids, a 30-meter lunar drill, and an advanced Apollo Lunar Surface Equipment Package.

Removal of the ascent engine increases the habitable volume of the cabin. Special hammocks provide more comfortable sleeping quarters for the astronauts. An airlock, attached to the forward hatch, serves as an EMU station and eliminates the need to depressurize the cabin before lunar surface egress.

LM TRUCK

This alternate LM Truck is a modification wherein all ascent stage components needed for unmanned landing are integrated into the descent stage. The vacated ascent stage volume can be filled with a 9,000-pound payload.

The representative payload shown consists of 760-cubic-foot fixed crew living quarters, Lunar Roving Vehicle, crew provisions for as long as 14 days, and a 4,800-pound, 550-cubic-foot scientific cargo.

A Truck landing represents only half of a dual-launch mission. After the payload arrives successfully on the moon, a second earth launch of a LM Taxi dispatches astronauts who use the life-support and scientific equipment in the LM Shelter in performing their duties during long-duration lunar explorations.

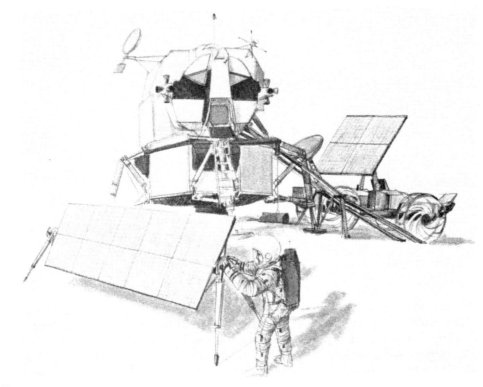

R-143

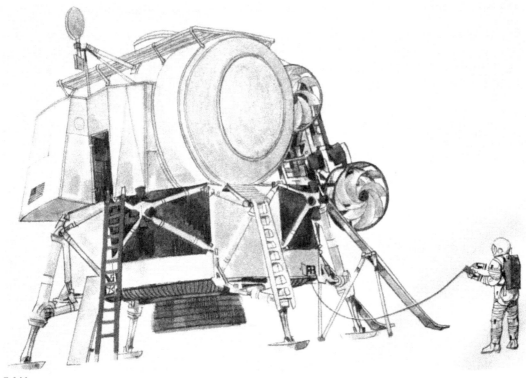

R-144

LUNAR BASE MODULE

The Lunar Base Module (LBM) provides the largest habitable volume for lunar surface operation; it retains the basic LM ascent stage structure. The ascent propulsion system has been removed, and the cabin enlarged to 450 cubic feet. The LBM, like the LM Shelter, is designed to land unmanned on the lunar surface, remain stored for 60 days, then support two men for as long as 14 days. Two beds provide maximum comfort for sleep. The LBM is used in conjunction with the LM Taxi in the dual mission mode.

Cabin volume is increased by enlarging the midsection diameter, moving the rear bulkhead back against the aft equipment rack, and removing the ascent propellant and helium tanks.

The Environmental Control Subsystem water sublimator has been replaced by a radiator, and solar panels have been added to the Electrical Power Subsystem. Both of these changes were made to support the 14-day mission with a minimum-weight vehicle.

GRUMMAN

MOBILITY AIDS

Flying and roving vehicles are being developed to support advanced lunar surface scientific exploration. Typical mobility aids that can be carried on the LM derivatives are:

Lunar Roving Vehicle (LRV)

Capable of supporting 1000 pounds, including two astronauts, their life support equipment, 200 pounds of tools, scientific equipment and lunar soil and rock samples; the LRV is small enough to be carried on a single-launch mission. Although its radius of operation will be 3 nautical miles, allowing crewmen to safely walk back to their landing craft in an emergency, the LRV will travel more than 20 miles during its lunar traverses.

The basic vehicle weighing less than 500 pounds, is about 10 feet long, 6 feet wide, 45 inches high, is battery powered, and is propelled by electric motors contained in each wheel.

Lunar Flying Vehicle (LFV)

This vehicle can be carried on a single-launch mission. It weighs 180 pounds dry and carries 300 pounds of the same propellant used by the LM descent engine; therefore, LM residual propellant can be transferred to the LFV after landing. The LFV can carry one man with approximately 370 pounds of scientific equipment or, on a rescue mission, two men without the equipment.

R-146

Dual-Mode Lunar Roving Vehicle (DLRV)

The DLRV is a 1,000-pound vehicle, including 350 pounds of scientific equipment; it has a growth capability to 1,750 pounds of which 750 pounds is scientific equipment. In the manned mode, the DLRV has a 6-nautical-mile radius of operation; in the unmanned mode, controlled remotely from earth, it can traverse more than 600 nautical miles. The DLRV would be carried on an Extended LM or on an unmanned logistic spacecraft such as the LM Shelter or Truck.

R-149

LM/STELLAR ATM

One of the purposes of the Apollo Telescope Mount (ATM) mission is to evaluate performance essential to the development of advanced manned orbiting solar and stellar observation systems. The experience gained on the ATM mission may be applied to other observatory missions by replacing the solar telescope with a large-aperture steller telescope. The Stellar ATM could then be used to collect scientific data on celestial objects in the ultraviolet spectrum. Such a configuration could be operated in a manned or man-attended mode of operation.

Periods of unmanned free flight away from the main orbital assembly would fall into the category of man-attended operation; such operation may be desirable because of the longer stabilization periods and higher pointing accuracy required with stellar targets. Such a configuration is shown in the artist's rendering. The solar array depicted is gimbaled to permit orientation with the sun, independent of the line of sight of the experiment package. The LM guidance, control, and propulsive capability, would be of particular use in this mode of operation.

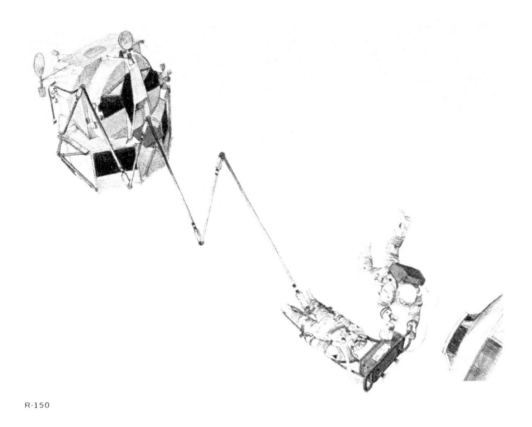

R-150

RESCUE LM

The LM's highly efficient Main Propulsion Subsystem, combined with the fully redundant Reaction Control Subsystem and versatile guidance and navigation capability, offers an in-orbit maneuvering, manned vehicle of unique capability. A LM with very limited modifications will permit application of this capability to effect in-orbit interception of, and rendezvous with, other space vehicles for such purposes as personnel rescue (as shown) and spacecraft inspection or repair.

INDEX

GRUMMAN

NASA
PROJECT
GEMINI

FAMILIARIZATION
MANUAL
Manned Satellite Capsule

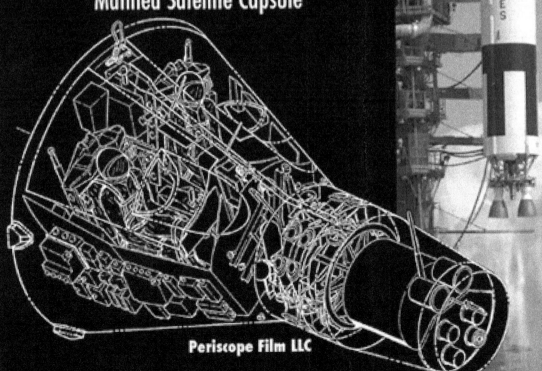

Periscope Film LLC

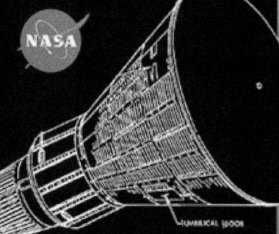

LMA 790-1

PROJECT APOLLO

lem
LUNAR EXCURSION MODULE

NOW AVAILABLE!

FIRST MANNED LUNAR LANDING
FAMILIARIZATION MANUAL

GRUMMAN AIRCRAFT ENGINEERING CORPORATION • BETHPAGE, L. I., N. Y.

CPSIA information can be obtained
at www.ICGtesting.com
Printed in the USA
BVHW051940240120
570324BV00004B/148

9 781937 684983